RICHES OF THE RAIN FOR

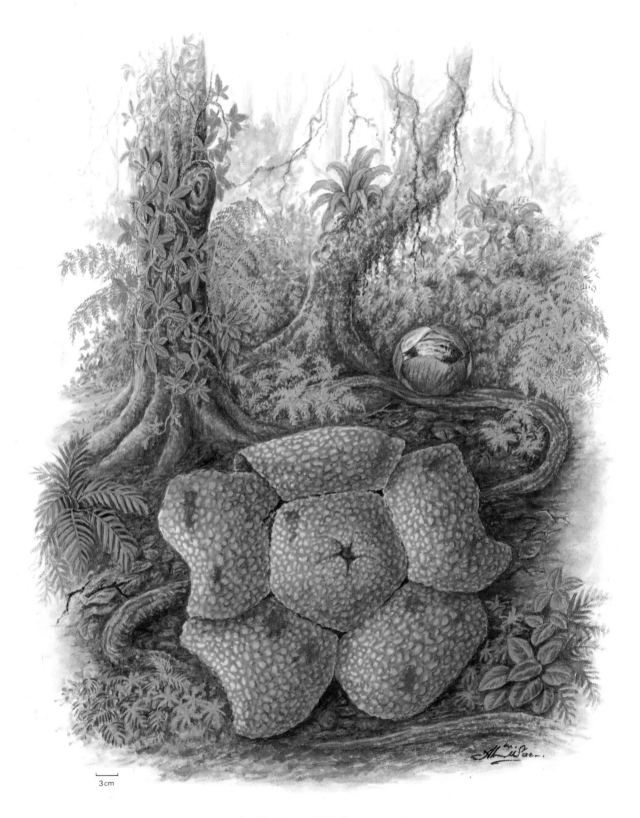

3 cm

Rafflesia arnoldii (*bunga patma*)

RICHES
OF THE
RAIN FOREST

AN INTRODUCTION TO THE
TREES AND FRUITS OF THE INDONESIAN
AND MALAYSIAN RAIN FORESTS

W. VEEVERS-CARTER

Illustrated by
MOHAMED ANWAR

SINGAPORE
OXFORD UNIVERSITY PRESS
OXFORD NEW YORK

Oxford University Press

Oxford New York Toronto
Delhi Bombay Calcutta Madras Karachi
Kuala Lumpur Singapore Hong Kong Tokyo
Nairobi Dar es Salaam Cape Town
Melbourne Auckland
and associated companies in
Berlin Ibadan

Oxford is a trade mark of Oxford University Press

© Oxford University Press Pte. Ltd. 1984

First published 1984
First issued as an Oxford University Press paperback 1991
Second impression 1992

ISBN 0 19 588989 4

Printed in Singapore by Kim Hup Lee Printing Co. Pte. Ltd.
Published by Oxford University Press Pte. Ltd.,
Unit 221, Ubi Avenue 4, Singapore 1440

To the memory of Dr Marius Jacobs

Foreword

THE tropical rain forests of the world provide more than just beautiful scenery and fresh unpolluted air. Neither are they just a home for wild animals or an interesting collection of vegetation. They are a source of abundant natural resources, a natural genetic bank and a living apothecary. Furthermore, the ecological role of forests in soil, water and atmospheric conservation is absolutely critical not only to the countries in which these forests exist, but to the overall health of the whole planet.

Truly tropical rain forests are a priceless natural heritage. Yet, humankind had not treated this priceless heritage with the loving care that it deserves. What has taken hundreds of millions of years to evolve has been terribly damaged, and is being irrevocably destroyed, at rates that could herald the destruction of humankind itself.

Thankfully, there is today, as never before, a growing debate on the fate of the tropical rain forests among governments, environmentalists, world organizations and the man in the street. But as with all hot political debates, there is considerable effort needed to separate the fancies from the facts.

Riches of the Rain Forest, when it was first published in 1984, served an important role in engendering an interest amongst laymen and amateur naturalists, especially through its beautiful illustrations. This revised version is indeed timely. I hope more people all over the world will read it and think a bit more about the need to conserve our natural heritage.

April 1992   TAN SRI DATO' MOHD KHIR JOHARI
President
The World Wide Fund for Nature (WWF) Malaysia

Preface

EIGHT years have passed since the first edition of this book, eight years of siege as far as the world's rain forests are concerned. After the El Nino-induced drought of 1983, an estimated 3.5 million hectares of forest burned in East Kalimantan alone. One would think that ever-wet forest could not burn, but drought conditions combined with poorly practised selective logging ensured plenty of dead wood to feed the fires, and in the abnormally dry peat forests the fire stayed underground. Record high surface temperatures maintained by the blanket of smoke haze made the situation worse. Nor did reserved forest escape, though investigating scientists found that the prospects for recovery in it were twice as good. Fire has since continued to destroy forest of all kinds in Malesia during the drier months (June to October), though on a smaller scale, while of course the logging for commercial gain continues. In Sabah and Sarawak, both under severe logging pressure, rivers have been silted up, soil eroded, fish killed, flooding caused and the forest peoples, who bear the brunt of this exploitation, displaced and deprived of their livelihood. 'Reafforestation', by governments or logging companies, remains problematic. It would be more honest to admit that man cannot re-create rain forest and so eliminate statistical deceit. Meanwhile, industrialized countries, with Japan heading the list, continue to import, from willing sellers, rain forest timber: short-term revenue, ultimate impoverishment.

It would seem, these days, that even rain forest with protected status is little safer from human interference than from fire. Sabah has demoted her National Parks to 'Sabah Parks' ostensibly to 'return them to the people' (who cannot protect them from the logging companies). The Trans-Sumatra Highway has been pushed through the last barriers of 'jungle', that is, through the middle of the Gunung Leuser National Park, once Sumatra's largest. This diminishes its conservation value, and opens the roadside forest (as in Brazil) to 'spontaneous' settlement. The logging of the Philippines' last undisturbed forests on the island of

Palawan continues in spite of strong (but not strong enough) local protest. Rain forest continues to be sacrificed to Indonesia's transmigration policy in spite of international criticism and the dissatisfaction of many of the transmigrants. The latest plan (mentioned in *Oryx*, January 1992) is to settle 40,000 Javanese on the island of Siburut, one of the Mentawai Islands, where up to now the native islanders have maintained a precarious balance with nature (see Epilogue, pp. 87 ff.) The Mentawai Islands, incidentally, were declared a Biosphere Reserve in 1981.

So, is there any good news? Just as logging exposes the forest floor, so the forest dwellers have become media-aware. It has taken time, but now they protest visibly. There are pictures in the world's press; the destruction is documented. In an attempt to counterbalance the inexorable economics of timber revenues, the Debt for Nature idea has at last been put into practice in a few places. The Philippine Haribon Society was created with such money, though it is not yet able to prevail against vested logging interests. There are now more universities, and more journals, specializing in tropical botany, ecology and biology than ever before; Aberdeen, since 1989, even offers an *under*graduate degree course in *tropical* environmental science. Increased academic interest means there are more and more trained people including students from rain forest countries working in this field, more educated protest, more knowledge gained. Though what they learn cannot keep pace with what is destroyed, every new discovery is more ammunition for using the riches of the rain forest in a sustainable way. And many more books have been published on the world's rain forests since 1984 (see the revised Bibliography). No longer the preserve, as far as foreigners are concerned, of the eccentric botanist, the avid orchid collector or the lumberjack, the rain forests have been popularized—on coffee tables, in 'the year of', in schools, in songs, in newspapers, even in ice cream: Americans, at least, have been prodded with the ethics behind the creation of Ben and Jerry's ecologically pure Rain Forest Crunch, and this was considered good advertising copy. Silly? I don't think so. I think these are all, including the ice cream, signs of a hopeful trend. For example, one of the most powerful organizations in the world has just taken a decisive step in the right direction: as of July 1992, the World Bank will no longer fund logging projects in tropical rain forests.

Guernsey   W. VEEVERS-CARTER
1992

Acknowledgements

THE author would like to acknowledge her debt to all the authors listed in the bibliography for ideas, inspiration and guidance. A text cluttered with footnotes or with academic-sounding references such as 'Corner, 1964' or 'Raven, 1979' is difficult to read, however correct it is to itemize each particular debt! But special thanks must go to the helpful librarians of the Kew and Bogor Herbaria, to Dr John Dransfield of Kew who critically reviewed the chapter on 'The Rattans', to Dr Timothy Whitmore, who did the same for that on 'The Agathis', to Dr Nicholas Cooling, for his help with Chapter 15 on 'Pinus Merkusii' and, most of all, to Dr Marius Jacobs of the Rijksherbarium, Leiden, without whose special inspiration in the beginning, and year-long help and encouragement thereafter, this book would never have been written at all.

As the book reached final proofs, the sad news arrived of Dr Jacobs' sudden death. He will be sorely missed.

The contribution of the artist, Mohamed Anwar of the Bogor Herbarium, needs no encomiums from me. I am particularly grateful for his skilled application, in spite of a heavy work load, to the illustrations where 'botany' has been tempered by 'art'.

Indonesia
1981

Contents

Figures

Plates

A Note on the (Confusing) Name of 'Malesia'

THIRTY years ago, when I was in school, we learned about two Asias: continental Asia, and the Asia of the archipelagoes off her eastern shores. These were, more specifically, the islands off her south-eastern shores: Japan was excluded as too far north, too different, too much itself to be part of anything. The south-eastern islands had one thing in common: they were settled, however lightly, by people of the Malay 'race', speaking a Malay language. Thus the name: Malaysia. And it also made sense to a child as well as ethnographically that the almost-island, or peninsula of Malaya was included. Luckily, this made sense to the botanists as well: just north of Malaya there is a clear separation, which they call a 'disjunct', between mainland Asiatic flora and that of Malaya and the island groups further east. But exactly where to the east Malaysia stopped was less clear, and depended on one's scientific persuasion. The ethnographers usually included the Moluccan islands, New Guinea and the islands east of it in other island-generalizations like 'Melanesia' or 'Micronesia', but the boundary was vague. Malaysia was not, after all, a political entity.

On 16 September 1963, however, the Federation of Malaya, newly enlarged by the addition of Sabah and Sarawak, took the name and made it political, and left the ethnographers and botanists to cast about for another. The ethnographers gave up the name altogether; it had never been of that much use. The botanists, concerning themselves with families of plants rather than of men, simply changed the spelling. By a nice combination of politics and orthography, they were freed from using a term that had another meaning to other disciplines, and they happily ignored, as they always had, all cultural, racial and political boundaries as irrelevant. *Malesia* is *their* word, and to them it extends from the Malay Peninsula, the Philippines, Borneo and all the Indonesian islands right through the whole of New Guinea to its eastern islands, in the Bismarck Archipelago. This is the botanical Malesia.

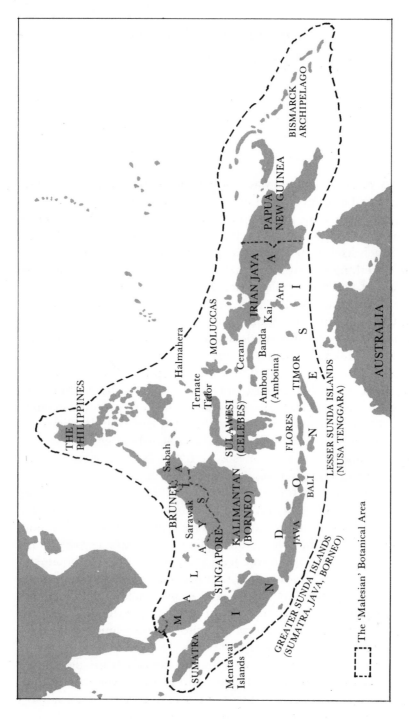

South-East Asia showing botanical 'Malesia'

1
The Dipterocarps

THE big family of trees known as the 'dipterocarps' are the giants of the South-East Asian forests and the dominant family in them. It is their presence that gives the lowland rain forests of Malaysia and the Greater Sunda Islands, the tallest rain forests in the world, their characteristic height. From the ground, their massive and often buttressed stems stretch upward out of sight, clean-boled because of their self-pruning habit, until they branch out into the characteristic cauliflower-like crowns which dominate the upper canopy. Combine the trunk's perfect shape with the fact that the wood is both hard and light and that most species will float, and it is easy to see why the dipterocarps are almost ideal from a lumber jack's or timber merchant's point of view—and therefore why they are so valuable. In South-East Asia they are referred to as dipterocarps generally or by the common local names of *keruing* and *meranti*. Trees from this family constitute the bulk of the timber exports, and when people talk about 'the great hardwoods' of these forests, it is this family they mean.

The name 'dipterocarp' comes from the Greek for 'two-winged seed', *dis pteron karpos*, although some *Dipterocarpaceae* seeds have three or five. If the seeds have 'wings', does this mean that they are dispersed by the wind? This would be a very unusual quality in a rain forest tree, indeed. Wind plays a much smaller role in the pollination of flowers and the dispersal of seeds in the tropics than it does in the temperate regions, and near the equator, winds are especially weak: the typhoons, so characteristic of the sub-temperate regions are rare here. But the term 'winged' is deceptive. Dipterocarp seeds are more like badminton shuttlecocks, the wings giving direction to the seed rather than lift. The seeds are not long-lived and are very sensitive to degrees of temperature and moistness; they need to find earth quickly in which to germinate and to avoid being caught up in branches when descending from the great height at which they have been produced.

10 m

Fig. 1   *Dipterocarpus palembanica* (keruing)

In the whole family there are about 500 species; 380 of them are found only in the Malesian area and most of these are confined to the lowland or low montane forests, the majority growing between sea-level and 800 m above it. But where they do grow, they are extraordinarily plentiful. Half the trees whose crowns can be seen from an aeroplane—the *emergents*, as they are called—are dipterocarps. No one species among them dominates a particular section of forest, however. Instead, individuals of any one species will be scattered thinly over a large area, so much so that in a single hectare of forest one might find only four trees of the same species. The finder would also have to be an expert: dipterocarps look much alike, and are usually identified by precise and detailed examination of the bark or leaves, or, if the wood is cut, by the pattern of the resin canals. Identification by the flowers or seeds is of course the best way, if you are lucky enough to find any. Fig. 2 shows the five-winged seeds of the *meranti, Shorea leprosula*.

'Lucky enough?' When the dipterocarps are so common? It comes as a surprise to most people to know that it is sometimes sixty years before a mature tree flowers for the first time, and that then, having flowered, it is likely to take a rest for three, seven or even eleven-odd years before flowering again—or perhaps longer! A botanist studying the dipterocarps has to be very patient and persistent if he is not to work only with what the botanists call 'sterile material', that is, with leaves and bark and the appearance of the wood under the microscope, rather than with flowers and seeds.

Furthermore, in the ever-wet forests no species of dipterocarp flowers regularly, even by its own odd standards. Some species seem to flower in response to slight changes in the humidity or the temperature, a dry period after a long wet one probably being the trigger, as it is for many of the rain forest trees bearing edible fruits; even thus triggered, flowering will only occur after a long period of non-flowering. But then, when some start to flower, each species in the same forest which is ready to flower will do so in succession, one group after another. They may, as a family, be responding to the same stimulus, but they never 'mix'. Cross-pollination or inter-breeding between different species does not occur—cannot occur—and thus each one retains its homogeneous and distinct identity.

Since flowering does occur so rarely, the years when it happens are known to those who care as 'dipterocarp years', and are high points for advancing the study of the family. They are also eagerly

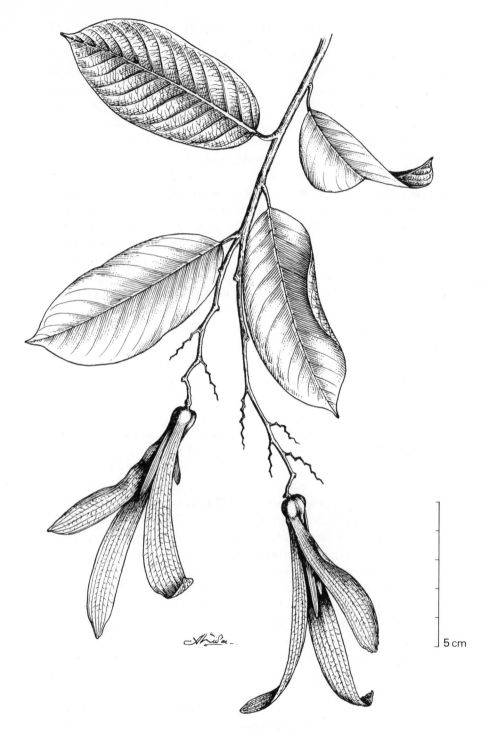

Fig. 2 *Shorea leprosula* (*meranti*)

awaited by local people who use the seeds of some *Shorea* species which grow along the rivers in Borneo for their high oil content (illipe nuts). But no animal life, tropical botanists included, actually counts on dipterocarp flowerings or fruitings for assistance in survival. If there were to be any relationship between a dipterocarp and, say, an insect, bird or other creature of the high canopy, what would that insect or bird live on in all the sterile years? Perhaps these giants of the forest have, in their evolution over millions of years, 'responded' to the generally poor soils in which they grow by avoiding 'help'. No help from the wind, none from insects or birds or any other animal, rare flowering, no cross-breeding: all of which has meant a very slow rate of increase. It is a sort of arboreal birth control which has resulted in a steady but very slow spread on land, perhaps as little as 1 km every hundred years. Water is also a barrier to most dipterocarps. Within Borneo, many species are found on one side of a river and not on the other, and most of them stop at the Straits of Makassar, between Borneo and Sulawesi.[1]

In view of all this, their success as a family is amazing. And to find that such gigantic trees will grow so commonly on such poor, heavily leached soil is even stranger. Since the average tree size seems quite unrelated to soil fertility, it is clear that in some way dipterocarp forests are largely independent of soil conditions. To some extent, this is true of all rain forest trees. The rapid recycling of rotting vegetable matter, at which insects, fungi and bacteria are all hard at work, is the process by which nutrients are liberated for quick absorption by plant roots. But valuable minerals and salts deposited on any soil subject to such high rainfall are almost immediately leached out, even when the soil is under protective forest cover; for agricultural purposes, such soils, unless they have their mineral content resupplied through volcanic action, are practically useless, or soon become so. Yet in non-volcanic Borneo, where the greatest number of dipterocarp species are to be found, some of the finest stands are found growing on the very poorest soils of all. It seems impossible. One can only guess that part of the explanation may lie in the special relationship dipterocarp roots

[1]The erratic flowering not only means slow dispersal and difficulties for eager botanists and illipe nut gatherers. It also makes nonsense of 'cutting cycles', 'minimum girth', and other regulatory controls over the timber concessionaires. An apparently mature tree may never have flowered and thus borne seed. Under the thirty-five year cutting cycle currently in vogue, many trees fall 'without issue', eliminating their genetic material from the forest forever.

have with certain fungi, a condition known as mycorrhiza, or 'fungus-root'. Pines and beeches also share this characteristic, which is one reason they are able to grow on such poor, exposed sites as landslips or mountain screes.

Not only do the poor soils of Borneo carry fine dipterocarps, we must also conclude, by the number of different species recorded there, 262 out of the 380 listed for the Malesian area according to the latest count, that it is in Borneo that dipterocarps have found the environmental conditions most suited to their development; more particularly so since Borneo is not their land of origin. The evidence is that this family has spread, over millions of years, from Africa via India, which carried the stock while drifting northwards to its present position (wedged against the Himalayas) some 40 million years ago. From India, the family spread eastward, but it was only, once they had arrived in the lands of the Sunda Shelf—in Malaya, Sumatra, Borneo, Java and the Philippines—that they proliferated and differentiated in their slow way, for about 35 million years, into the hundreds of species that we know today.

Altogether it is a very interesting family of trees with, now, few relatives left outside this area. In Sumatra and Java dipterocarps in commercial numbers disappeared years ago, and in lowland Malaysia and the Philippines they are seriously overexploited. Borneo has taken over as the main focus of timber exploitation in South-East Asia, but how much longer will these forests last? At the present rate of exploitation, they will be depleted by the early years of the twenty-first century. Indonesia's most pressing conservation need is to establish large enough lowland dipterocarp reserves now, before it is too late.

1m

Fig. 3 *Ficus benjamina* (*waringin*) (a strangling fig)

2

The Figs

MANY are the forest-dwelling animals—insects, birds, civets, monkeys, apes and bears—who owe an element of their survival, sometimes a very large element, to the figs of the Malesian forests. Whereas the dipterocarps seem to have avoided dependence on any other form of life except perhaps on the fungi in their roots, figs have gladly accepted the assistance of a multitude of creatures, and given much to them in return. Even the human beings who only look upon trees as so many cubic metres of timber have cause to be grateful to the humble fig, in this case, to the so-called 'strangling fig'. Why? Because this particular type of fig is often the first of the rain forest trees to colonize the tangle of secondary growth of short-lived pioneer plants and grasses.

Seeds of strangling figs are dropped by birds or monkeys high on the branches of trees they pass; here they germinate, then either send a root down the trunk of the host tree or grow an aerial root to the ground from the branch. Once this root is established, the fig begins to grow, expanding its first stem and producing leaves upward while simultaneously establishing more roots downward, a growth system which eventually surrounds the stem of its host. Finally, when the fig has spread its large crown above that of its host tree, the host tree dies, leaving the fig standing on its basketwork of roots as a huge and independent tree, providing in its turn the shade without which the seeds of the true rain forest cannot germinate. And yet these 'stranglers' seem to be the bane of professional foresters and are often deliberately killed in the cause of 'habitat improvement'; a strange perversion of the biological facts.

Figs have many other growth forms as well. They appear in the forests as climbers, as creepers, as epiphytes or as quite ordinary trees with, however, some quite extraordinary ways of fruiting and methods by which the fruit it set.

One of the most striking features of the rain forest is the way some of the trees in several different families bear fruit not on the ends of twigs but on the thicker branches or even hanging singly or in clusters on the trunk itself. This arrangement, known as 'cauliflory', is both interesting and puzzling. In no other environment do you see the leaves and branches of a tree high above your head while the flowers and fruits are on the stems in front of you. How and why has this strange habit evolved? Does it have survival value for the tree? It certainly makes the fruit more accessible to some animals, particularly larger ones, but puts others like monkeys and birds to the trouble of leaving the relative safety of the upper canopy. It does frustrate the orang-utan's tendency to break off whole branches of a tree merely to eat the fruit, but this seems an unlikely explanation from an evolutionary point of view. A few fig species even fruit on long horizontal runners under the soil: sometimes you bend down to pick up a fig lying on the litter of the forest floor only to find it attached to a nearby and apparently sterile tree. What purpose does this serve? As the botanist E. J. H. Corner suggests, in *The Life of Plants*, these unusual ways of fruiting may simply be primitive characteristics, or evolutionary sidelines, which figs, like durians, *langsat*, jackfruit and others, have not yet been forced, through selective evolutionary pressure, to discard. Corner's interesting theories on the forms of primitive trees and fruits are discussed in Chapter 8.

But the tendency to cauliflory and ground-fruiting, though remarkable, is not the most fascinating aspect of the fig family. This is reserved for the methods by which the flowers are pollinated and the fruits set. Each species of fig in the forest is absolutely dependent for this service on one particular species of fig wasp; conversely, if the female of this wasp species cannot find her particular kind of fig to lay her eggs in, she and all her eggs die, and the unpollinated flowers will not develop into fruit. In the Bogor Botanic Gardens, for instance, only the figs which are native to West Java produce fertile seeds, while the others remain infertile because of the absence of their species of wasp. Both plant and insect, in other words, are highly specialized and specific to each other: without its wasp, the tree is sterile; without their trees, the wasps would become extinct.

It may occur to the reader that he or she has never seen a fig tree in flower. The fruits are familiar, but the flowers? They are actually inside the fruit or, looked at another way, the fruits are the mature flowers. Each fig is an inflorescence or collection of reproductive units (flowers) which, instead of being displayed as is usual in

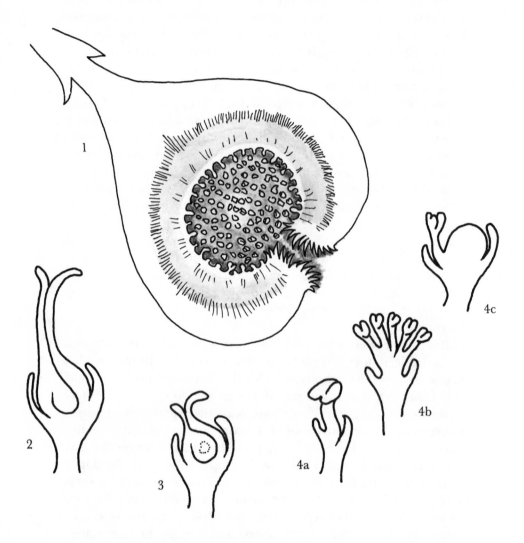

Fig. 4   A fig inflorescence and types of flower, in section. (1) The fruit or syconium; (2) female flower (long style); (3) male flower (short style) showing wasp egg (dotted circle) in position; (4a) and (4b) two types of male flower; (4c) male flower with gall ovary (*Ficus variegata*). Only monoecious figs have male, female and gall flowers in one syconium.

Source: After Hrdy and Bennett; 'The Fig Connection', *Harvard Magazine*, September-October 1979; illustrated by S. Landry.

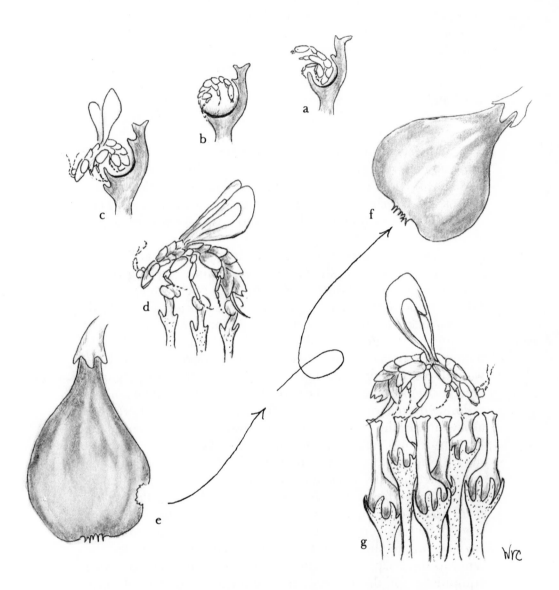

Fig. 5 A fig wasp's life cycle. (a) Male wasp leaves gall flower first;
(b) inseminates female still unhatched; (c) impregnated female emerges
and (d) collects pollen from male flowers; (e) female leaves fig of birth and
flies to another ripening fig of same species (f); (g) fig wasp searches for
gall flowers in which to lay eggs, meanwhile fertilizing female flowers.

Source: See Fig. 4.

plants, is turned in upon itself and, in addition to being invisible, is well-protected by a mass of interlocking scales or bracts which bar its single entrance (Fig. 4). No casual pollinators will enter here, but only the female wasp specific to that particular fig. It is sometimes possible to infer that a fig is in flower, however, particularly if the fig tree stands apart from other trees, by the number of swifts darting over and around the tree, trying to snap up the wasps as they emerge. This is a common sight around the *waringin*, *Ficus benjamina*, a popular (and holy) village tree.

Each species of fig produces three types of flower: male, female and gall flowers. The male produces pollen, the female sets seed and the gall flower, which is actually a sterile or seedless female flower, acts as an incubator for the next generation of fig wasps. The continuing production of both fruit and wasps is a masterpiece of timing and design.

This is how it works. Fig wasp maturation times and flower opening times are carefully staggered, both within each fig (which can contain more than one type of flower) and between fig trees of the same species (see Fig. 5). Of the wasps maturing in the gall flowers, the males are the first to emerge. Blind and wingless, they find and inseminate the female wasps while the latter are still inside their flowers; then, their single function performed, they die— without ever having left the fig. Now the females, already pregnant, emerge. At the same time, the male flowers ripen and their anthers become covered with pollen. Since male flowers and gall flowers are always inside the same fig, and are always positioned near the 'mouth' or single opening, before the female wasp can emerge from the fig, she must pass over the male flowers, thereby collecting pollen on her body. Once out, she then flies (if not snapped up by a swift) to another tree of the same species whose fig flowers are ready to mature. She enters a fig and starts to probe the various types of flower in order to find the gall flowers in which to lay her eggs, one egg to one flower. From the top, all these flowers feel the same to the wasp; she has to test each one with her egg-laying equipment, for the flowers are anatomically different. Female flowers have long, curved styles and her egg-placer or ovipositor cannot reach down these styles to the ovary. This feels 'wrong', and frustrates egg-laying, but she has meanwhile fertilized this flower with the pollen she carries. Gall flowers, however, have short styles; in these her ovipositor easily reaches the ovary and egg-laying is accomplished.

Male flowers are not a problem since they have opened before she herself has hatched and now she is inside another fig on another tree operating to a different time schedule. If, however, she is a wasp specific to the kind of fig which keeps its female flowers sequestered in separate figs and she happens to enter such a fig, the unhappy wasp will pollinate all of the flowers but not be able to lay a single egg before she dies: her life span is too short to accomplish more. Bad for the wasp generations, but good for the consumer: gall figs, which look just as ripe and juicy from the outside, are not at all nice to eat since for every seed there is instead a developing wasp, and the figs of stranglers, which always contain all three types of flower, are therefore guaranteed to be the least appetizing from the human point of view.

Once the fig flowers and the wasps have played their vital roles for each other and the 'fruit' has set, the harvest is always generous. Figs mature throughout the year, and the delicious and nourishing fruit attracts all the animals of the forest. For birds, from barbets to hornbills; for gibbons, leaf monkeys and macaques and even the large orang-utan, as well as for civets and squirrels, figs are an important part of the diet. Almost a quarter of all the fruits an orang-utan eats are figs and they readily discover new sources of supply by observing the purposeful flight direction of the noisy hornbills, who reveal to all the forest that a new fig tree is in fruit.

This plentiful food supply not only enables many animals to survive but also serves, literally through them, to disperse the small fig seeds and help them to germinate by dropping them along with a supply of manure. Generous harvest, generous propagation: second only to plants of the *Eugenia* genus (see Chapter 5), the fig genus, *Ficus*, is the largest genus of woody plants in the Malesian area, with about 100 species in Malaya, 109 in Borneo and 122 in New Guinea. Common and, unlike many other rain forest trees, usually easy to recognize: the stranglers from their distinctive way of growing, and the others from their smooth grey trunks and thick simple leaves which turn a characteristic yellow on fading, and, being shed often, are usually found lying in great numbers beneath the trees.

3
The Rattans

RATTANS, whose amazingly versatile stems are made into so many familiar products the world over, are the climbing palms of the tropical rain forests. Two-thirds of all the palms found in such forests are rattans and, conversely, the rain forest is where they thrive best. Some few species are cultivated at the edge of the forests, and there are several others which grow well in secondary vegetation or along river banks, particularly in Sulawesi. But the main stock of rattans and most of the desirable species exist and can only exist in the 'primary' or undisturbed rain forest, and can only regenerate naturally under true rain forest conditions.

Rattans are, furthermore, most typical of Malesian forests. Apart from the few found in Africa, most species (480 altogether) grow in the Malesian area, and of these, the greatest number on the islands of the Sunda Shelf. Except for one species, none of the Sunda rattans is found further east than eastern Borneo (Sabah and East Kalimantan), but many others grow in Sulawesi, where they have great commercial importance as an export, and there are still other species in the Philippines, the Moluccas and New Guinea.

The fact that there are so many different rattans is the reason for their popularity and, therefore, for their commercial value. It is as if all sizes of rope or cord, from hawsers to string, were grown free of charge, literally 'on trees', and were available to anyone willing to make the effort of collecting them! In thickness, rattans range from a diameter of 3 mm to almost 10 cm and in length, anywhere from 10 cm to 200 m, and these stems are of uniform thickness (or almost so) throughout their entire length, pliant yet strong, springy, durable, and smooth enough after processing to be used entire, as in the arms of a chair, or to be split easily into strips and woven in the ways with which we are all familiar. Apart from furniture, in South-East Asia rattans are used for making every imaginable sort of container, from baskets to fish traps to suitcases. They are also made into light, springy balls for certain games and twisted for ropes, and even used whole for walking sticks and umbrella handles, for which a rattan with many nodes close together is selected. The ease with which rattan can be worked when split, or steam-bent into permanent curves which still retain a certain resilience makes it the ideal material for all these articles. It often seems, in the South-East Asian region, that what cannot conveniently be made of rattan is made with bamboo, and vice versa, to such an extent that it is difficult to imagine living in this part of the world without both these useful groups of plants.

But as the forests decrease, so too does the supply of rattans, whereas most bamboos, which thrive on disturbed land and are readily cultivated, are not in danger. In recent years, we have already seen the effects of decreasing supplies in the rising prices for products made with rattan, and as more forest is cut down we can expect that certain species will become altogether unobtainable, and that the many people for whom the rattan trade constitutes a livelihood will have to find some other source of income. This includes not only the collectors, who are usually people living in or near the forest, but also the many middlemen and craftsmen involved in the trade. The continued supply of rattans is therefore one component of the economy of South-East Asian nations and should not be dismissed as the concern of the botanists alone.

Climbing plants have a variety of devices to help them on their way up into the canopy of the rain forest—special twigs growing rigidly at right angles or tendrils which twine either to the right or the left around any available object. Rattans are more aggressive: the rattan palm climbs by means of a long whip-like organ, armed with hundreds of reflexed or recurved thorns, which grows at the end of the leaf or leaf sheath (see Figs. 6 and 20 and Plate II). As this device waves about, it attaches itself to any convenient object, and the angle at which the thorns are set enables them to penetrate even further with any additional movement; wind or storm or the passage of animals only serves to anchor the rattan more firmly. The presence of many rattans in a forest is therefore a painful nuisance to any passing traveller! On jungle paths, the local people cut down the young plants automatically, since no cloth or skin or hide is immune to the clinging, gouging action of the thorns. As if this were not enough, in many species the stems themselves are thorny or bristly or are covered by prickly leaf sheaths which characteristically remain for a long time on the stems. By all of these means, the rattan will be firmly fixed in any position its climbing

Fig. 6   Daemonorops oblonga (rattan)

organ has originally won for it; all downward movement is strongly resisted, and harvesting, with few barbless parts on which to pull, is correspondingly difficult.[1]

When the rattan stem or cane has been collected, it must be coiled up, or cut into sections if its diameter is large, and carried out of the forest, often several days' journey away from the place of processing. It is important that it reaches the processor as soon as possible or it will begin to deteriorate. The big canes are particularly full of gums, resins and water, all of which must be removed if they are to be durable, by boiling them in either coconut or diesel oil. This process is known as *layang*. The largest canes are cleaned first by hand to remove the leaf sheaths (this is sometimes done in the forest); after boiling, they are stored in bundles until sold. The smaller diameter canes are either boiled or fumigated over burning sulphur, which bleaches at the same time as it preserves. They are then dried and sold whole, in coils by the kilo, or are split into 'skin' lengths (the outside) and 'core' lengths, either by hand or by machine.

Depending on the use for which the cane is intended, either whole or in strips, flexibility may be more important a quality than natural shine. In this case, while the cane is still whole, the silica-filled outer layer is removed by a process known as *runti*, that is, it is rubbed with sand, scraped with chain or, more sophisticatedly, pulled through a series of bamboo or ironwood rollers so placed that the cane is forced to bend back and forth in S-like curves, causing the silica layer to flake off. Shiny canes are used in floor mats, the major structural parts of furniture and woven seats or table tops; the more flexible ones for basketry, decorative or structurally necessary bindings and so on.

The role of the harvester in selecting the rattan species he picks is just as important as the role of the processor, however, because only mature stems can be preserved correctly and processed to give a good quality commercial product.

But against being harvested at all, rattans seem to have many natural defences. Besides their tenacious thorns and persistent leaf sheaths the form of which greatly assists both scientist and collector in telling one species of rattan from another, there are also many layers of leaf sheaths which protect the young growing point or

---

[1]This said, I was reminded by Dr Jacobs that 'an experienced botanist'—he qualified, of course, for the title—'in less than an hour's time is able to cut up and prepare all the essential parts (of a rattan) for subsequent study in the herbarium'.

heart of the rattan. As in the coconut and many other palms, this heart of tightly folded immature leaves is delicious eating; at this stage, the inside spines lie soft and flat and only contract into the formidable right-angle position as they stiffen and harden when exposed to air. But tribesmen, wild pigs and other browsers who want the 'cabbage' still have to get through the outer sheaths. Then, too, some species of rattan, the *Korthalsia* or *rotan merah*, for instance (see Chapter 13), are inhabited by ants which bite fiercely when disturbed, and many rattan fruits, which look like scaly beads similar to the fruits of the sago, raphia and salacca palms, have distinctly bitter-tasting flesh. Yet in spite of all these 'keep off' characteristics, elephants and pigs take a high toll of young hearts within their reach, men harvest the stems, and birds, civets, squirrels, monkeys and some local people all eat the fruits with apparent relish; an excellent dye can also be extracted from them.

The seed stage is in fact probably the most vulnerable point in the rattan's life cycle. If the seeds are not destroyed when the fruit is eaten, they fall on the forest floor, but then can take anything between one month and six to germinate. During this time, they can endure no drying out, and once they have germinated, need very precise soil requirements for their development. The literally hundreds of successful seedlings one sees in undisturbed rain forest is the best evidence there is for the ideal conditions for rattan that prevail there; and, as a corollary, where rattans continue to germinate successfully, you can be sure that the forest soils of that area are still in satisfactory condition. But only the complex interaction of bacteria, insects, fungi and the functions of other root systems can retain tropical soils in this condition. Once the forest cover is removed, these cease, and full exposure to hot sunlight and the full force of heavy rainfall would certainly ensure that most rattans could no longer grow, even if they were not also by nature climbers in need of the support of tall rain forest trees.

4

The Ironwood of Borneo and Sumatra

ULIN, or *belian*, the ironwood tree native to Borneo and Sumatra, is the hardest of all Malesian hardwoods. It comes from a 'hard' family, the *Lauraceae*, which contains many valuable timber species, and although Borneo ironwood, *Eusideroxylon zwageri*, is not the only wood so nicknamed in the world, it is the most iron-like known here, and well-known at that. Indeed it is difficult to travel anywhere in Borneo, where there are more ironwood trees left than in Sumatra, without encountering floors, jetties, house posts, steps and stairs, walkways, pestles and mortars, weapons and even canoe and *prahu* hulls made from this useful tree—though not the paddles (it is such dense wood they would sink immediately). This ironwood's extreme durability and imperviousness to insect attack well repays the extra labour of cutting and fixing it, or of hollowing it out. It is also used to make the very handsome shingles, (the *sirap ulin*, Figs. 7a and 7b) which roof all the most sophisticated and beautiful buildings in Indonesia. The wood splits easily so that shingle-making is not so difficult as the hardness of the wood might suggest. Again, durability is the main attraction, but the asymmetric pentagonal shape into which the shingles are cut contributes much to the stylish look of the buildings they cover.

In general, hard wood grows slowly, and Borneo ironwood very slowly indeed. There is little data on the exact age of known trees growing under forest conditions (there are some fine trees in the Bogor Botanic Garden), but Dr Jacobs reports that he saw considerable regrowth of *ulin* in an old pepper plantation near Samarinda, in East Kalimantan. The plantation had been abandoned thirty-three years before he saw it, yet none of the young ironwood had attained a diameter of over 12 cm; at this rate, a tree of 100 cm diameter must be over 200 years old at least. Amongst the new growth, incidentally, stood some old *ulin* stakes which had been used as supports for the pepper vines, still as sound as the day they were put in.

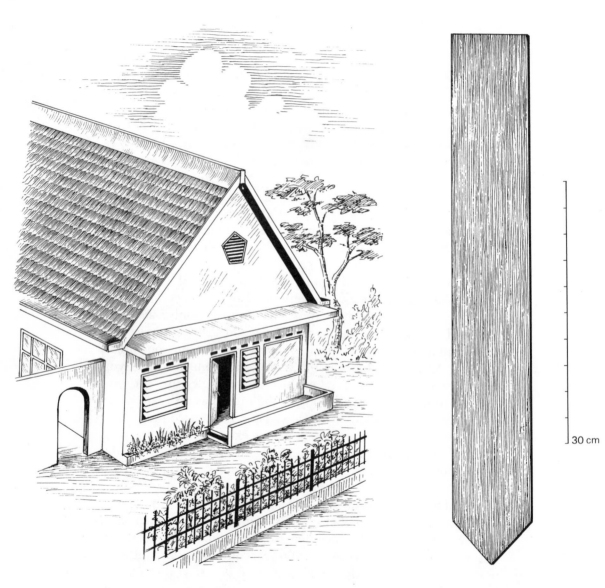

Fig. 7a A *sirap ulin* (ironwood shingle) roof

Fig. 7b A single *sirap ulin* shingle, showing its shape

30 cm

When they eventually mature, they are moderately tall trees, between 15 and 30 m high, with smooth boles, buttressed only slightly, and thin, dark brown scentless bark, the scentlessness being perhaps unusual since the *Lauraceae* family includes both cinnamon and the bay laurel. The leaves are large, long and pointed at the tip, smooth without marked veining and somewhat leathery. When the tree is in bloom, clusters of small flowers hang in panicles from the leaf axils. The fruit, however, is large and the seed one of the biggest in the world; a seed measuring 12 cm in length and 4 cm in diameter (see Fig. 8) is only about average. Seed sizes vary a lot from tree to tree, as a matter of fact, and even—a very remarkable fact—on the same tree! It is also an extremely well-protected seed, the nail-hard bony covering marked with curious long furrows overlying an only slightly less hard fibrous coating; within which is the seed. The coverings take three months to a year to rot away.

Once germination has begun, the size of the seed provides a storehouse of food for the new plant: *ulin* seedlings in the dark forest often grow to a metre or more before they begin to put out leaves. The ability to use its food resources solely to gain height is a useful one in the rain forest and—though not so marked as in *ulin*—is shared by many rain forest species with large seeds, such as durians, mangoes and avocadoes. (The avocado, a New World plant, is also a member of the *Lauraceae*.) But where *ulin* regeneration has been good, the presence of a lot of these snaky, leafless red-tipped seedlings standing nearly waist high gives a very odd look to the dark forest floor.

Forests dominated by ironwood, rare now, have a characteristically closed canopy at second storey level, which of course has a good deal to do with the seedlings' eager striving for height. Few trees in these forests break the 30-metre barrier to disturb the even blanket of green; this is not dipterocarp country, and in Borneo even such ubiquitous species as the tall 'king' tree, *Koompassia excelsa*, seem to get shaded out. Such forests are found only in the

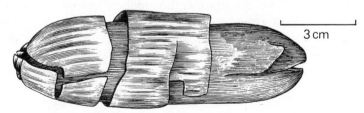

3 cm

Fig. 8   *Ulin* (*Eusideroxylon zwageri*) seed

lowlands, at never more than 600 m above sea-level and usually deep, well-drained sandy soil, and only rarely on clay, shale, marl or limestone. The thickest stands are, or were, found on flat land near rivers, with the nice explanation suggested that as the seeds are so heavy, they will always roll downhill from the tree, until they can go no further. Where such stands of ironwood remain relatively undisturbed, the natural regeneration is good, but the riverside sites they seem to prefer makes both legal and illegal exploitation easy. That ironwood sinks is not a problem: timber-cutters fell other trees as 'floaters' to hold the ironwood up for towing.

Since the commercial value of ironwood has been high for over a century, long before the dipterocarps were much thought of, the more approachable stands and single trees have long since been felled, and in West Kalimantan were depleted by 1925. In Central and East Kalimantan, virtually the only ironwood left grows within the boundaries of the nature reserves; but this is not the barrier to cutting it that it should be. Policing the maze of the extensive river systems of lowland Kalimantan is an almost impossible task.

That the seed can retain its viability for up to a year, if sealed within its slow-rotting covers, is a most unusual quality for a rain forest tree, and it suggests that the establishment of ironwood plantations ought to be possible. But although some experiments were made—one tree in the Economic Garden, Singapore, reached a height of 5 m before the garden was abandoned—none have succeeded. There are eight or nine trees in the Bogor Botanic Garden, including one exceptionally large and spreading example; though few of them bear seed, their mere presence is a tribute to the steady care this garden has received, through wars and all sorts of political emergencies, since its establishment in 1817. But in general, the slower growing the species, the more it will be at risk in garden conditions. Even if political or economic factors do not cause the garden to close, those in charge of a garden may, in the passage of time, lose interest in or the ability to care for a particular plant. In any case, the space and patience required to cultivate large stands of slow-growing hardwoods in a garden never have and probably never will exist, and the best policy is to protect a sufficiently large area of forest where they will grow and replace themselves naturally.

5
The Eugenia Genus

To those who have not been exposed to much science, the title of this chapter probably sounds like the name of a strange disease, or of a tribe from outer space. But in order to describe a plant or a group of plants, the use of a few Latin or Greek names is unavoidable. All of science as we know it today is imbued linguistically with one or another of these ancient languages, or, somewhat less recognizably, with words of Arabic origin. But words which sound suspiciously as though they are meant 'for scientists only' are really a great help. After all, common local names are only common to one language or, worse still, to one locality, whereas Latin or Greek names are used universally and therefore help rather than hinder understanding. So, too, do the various classifications into which the plants (or animals) are divided, first into big groups ('Classes', 'Orders') and then into smaller ones ('Families' and 'Genera', which is the Latin plural of 'Genus'), until finally you arrive at 'Species', which is the plant in front of you and those so like it in every respect that they cannot be further divided except into 'varieties' because of minor differences. Members of a species, including any 'sub-species' or varieties, can also cross-fertilize, but cannot breed outside this group.

Common names often reflect such groupings as genera or species. This is very much the case with the *Eugenia* species in Malaysia and Indonesia, where all the popular fruit trees are called *jambu*, for the genus, followed by another descriptive name for the species, as in *jambu air*, *jambu bol*, and so on, the scientific names following exactly the same order and idea: *Eugenia aquaea*, *Eugenia malaccensis*. The trees which do not produce comestible fruit, or at least fruits which are not commonly eaten by people, are also recognized as having affinities; these are known as *kelat*, from the astringent nature of their bark, which is used in tanning leather, dyeing sails or toughening and preserving fishing nets and lines. The word '*jambu*' means 'tuft' or 'tassel', and refers to the form of the

Eugenia flowers, showy on those with the best fruit, characteristic of the genus as a whole and very beautiful. When these somewhat heavy-looking trees are in bloom, the abundance of stamens in the flowers creates a pom-pom effect, so that the whole tree is covered in red, white or red and white balls of fluff, which are eventually replaced by the equally attractive pink or red fruits so familiar in this part of the world.

The *Eugenia* species are evergreen trees and have rather short trunks which separate into thick branches not far from the ground; this, combined with their dense foliage of dark green leathery leaves, is what makes them look 'heavy' when mature. The leaves of nearly all species hang downwards from the twig; they have drip-tip ends, which are a common adaptation of rain forest tree leaves enabling them to shed water quickly, and have, like other members of their botanical family, the *Myrtaceae*, the characteristic dot-sized oil glands on the leaves' underside (and elsewhere in their tissues), which make them smell so strongly when rubbed or crushed.

All told, there are about 700 species of *Eugenia* in the rain forests of Malesia, and since they occur from sea-level to the mountains, they are the most common, and commonly encountered, of all Malesian trees. As Corner says, in *Wayside Trees of Malaya*, 'a walk of half a mile in any part of the country, other than the limestone hills, will take one past half a dozen kinds'. Not even the wild figs, he points out, show more variety; the *Eugenia* species, like figs, are often found in secondary vegetation as well as in undisturbed rain forest because their seeds are also small and hard and easily distributed by birds, fruit bats, squirrels and other small animals.

Common as they may be, and therefore well known locally, the *jambu* and *kelat* trees have one famous relative of which everyone has heard: the clove tree, or *cengkeh*, *Eugenia aromatica*, the most aromatic of this aromatic family. Clove trees originally came from the Moluccas, specifically from the five islands off the west coast of Halmahera; the sultans of Ternate and of neighbouring Tidore, two of these islands, were once the rulers over most of the Moluccas even as far east as the coast of New Guinea, and their legendary wealth was based entirely on the value the Chinese, Portuguese and others set upon their 'odiferous nails' (the Chinese name for cloves), and on the nutmegs whose production and export from the Banda islands further east they controlled. For centuries, seventeen of them as a matter of fact, cloves and nutmegs were carried from their homeland to India and Europe, and many people became

2 cm

Fig. 9 *Eugenia aromatica* (*chengkeh* or *cengkeh*) clove

rich in the process, or lost their lives. So valuable a trade caused
many a small war between the rival nations trying to control it, and
subjected the islands themselves to a series of annexations and/or
monopolistic rulers from the sixteenth century until the middle of
the nineteenth, even though by this time both spices were being
grown in other countries. Ternate itself has had a particularly
stormy history, to which the several Portuguese, English and Dutch
fortresses dotting its coastline bear witness.

Cloves themselves are actually the immature, unopened flower
buds which are picked when they have just turned red; they are
dried and sold without further processing, or are made into clove
oil, which can also be distilled from the leaves by exposing them to
hot steam. This oil, eugenol, is much used in medicine as an
anaesthetic and an antiseptic, and indeed in their homeland both
cloves and nutmegs have little value as spices or condiments,
their use being entirely medicinal. If it were not for the fact
that Europeans found these strange medicines of the East both
preserved food and improved its taste, the Moluccan 'spices'
(= 'species') would never have been so much sought after, nor the
cause of so many minor wars.

When young, clove trees look quite unlike the majority of the
Eugenia species, though of course they conform on closer examination. Instead of drooping leaves of darkest green, young clove
leaves stand up perkily, light green and shiny except, in the newest
growth, when they are pink or red. The stems of the new shoots are
also red. Actually redness, either in the leaves and twigs or in the
fruit, sap or inner wood, is common in *Eugenia*, many of which
display colour in the young and sometimes the fading leaves, but
clove trees are particularly easy to recognize in that they 'flush'
continuously throughout the year, a pretty sight in the hillside
plantations.

Growing cloves in plantations is popular in Indonesia these days,
and the owners grow wealthy on the proceeds—a far cry from the
days when the Dutch sought to maintain a monopoly of the supply
and an artificially high price by restricting clove growing to the
island of Amboina (Ambon). It is only relatively recently that the
planting of cloves has actually been encouraged by the Indonesian
government: a sort of 'forbidden fruit' aura still seems to surround
this spice. During the last fifty years, however, their use within
Indonesia increased so greatly that Indonesia, once the homeland
of all the cloves in the world, has had to import them from

elsewhere. Until the revolution in Zanzibar in 1964 adversely affected the clove exports, most of the cloves used to flavour the popular *kretek* cigarettes were imported from there, where, remarkably enough, a superior variety of clove had developed. Nowadays there are so many Indonesians and so many of them who smoke that consumption by the *kretek* manufacturers accounts for nearly all local production, and in contrast to earlier times, the government is actively promoting the planting of clove trees both privately and in various land use and resettlement schemes.

The story of the ending of the Dutch monopoly by the theft of some clove seedlings, and of their consequent establishment in the French possessions in the Indian Ocean, is also the story of the spread of the nutmeg trees, and is told in the next chapter.

In New Guinea, the *Eugenia* species are much less common and are replaced in this sense by other members of the *Myrtaceae* family, especially by species of the typically Australian genera, *Eucalyptus* and *Melaleuca*. These trees are normally associated with a dry climate; indeed it is hard to think of Australia without thinking of the 'gum trees', quick-growing, hardy, excellent for windbreaks and good as timber. Nowadays they are grown all over the world where climate permits, even though they are said to 'mine' an area of its available water through their long tap roots, one of the reasons for their success in prevailingly dry areas. Not surprisingly, few *Eucalyptus* or *Melaleuca* species are adapted to life in the rain forest, but those few are very successful. *Eucalyptus delgupta*, for instance, a large handsome tree, is found as far west as Sulawesi and characterizes some of the forests there, and *Melaleuca leucodendron*, the paper-bark tree is very common in the New Guinea rain forests, especially in burnt-over swamp lands. All Myrtaceae have oil glands, which by distillation can yield valuable medicinal oils, but some species more than others. *Melaleuca leucodendron* is particularly well-known for this property in Indonesia, and is the source of the popular liniment, *minyak kayu putih*, commercially known as cajuput oil. The scientific name, *leucodendron*, is a direct translation of '*kayu putih*', or '*white wood*'.

The bark of *Melaleuca leucodendron*, the paper-bark, is also much used by local people, mostly as a caulking material for boats as it has the property of swelling when damp. The bark of three *Eugenia* species in Malaysia and Sumatra is used in a similar way, and as a styptic and absorbent dressing for wounds, staunching the flow of blood and having both a soothing and an antiseptic effect.

5 cm

Fig. 10   *Melaleuca leucodendron* (*kayu putih*), leaves and flowers

Fig. 11   A bottle of *minyak kayu putih*

Tea can be made from the leaves of another relative, *Leptos-permum*, common in the mountainous areas of Malesia, and another *Myrtacea*, of American origin, is the guava, called here *jambu batu*, a local recognition of its botanical similarities even though its Latin name, in this case, differs: *Psidium guajava*. *Pimenta officinalis*, allspice or pimento, is another *Eugenia*-in-disguise.

As fruit-bearing trees, as spices, as sources of tannin and bark with other useful properties, and as sources of medicinal and flavouring oils, the *Myrtaceae* are indeed a useful family of plants, of which the *Eugenia* genus is the most versatile and widespread. Here is a list of the commonly cultivated species, only a few out of the hundreds of relatives to be found in the rain forest.

Eugenia aquaea	*jambu air*
Eugenia jambos	*jambu mawar*
Eugenia malaccensis	*jambu bol*
Eugenia cumini	*jambu jambolan*
Eugenia javanica	*jambu air rhio*
*Eugenia aromatica**	*cengkeh*
Eugenia uniflora	*asam selong*
Eugenia pachyphylla	*kelat*
Eugenia palembanica	*kelat semak*
Eugenia cymosa	*kelat gelam*
Eugenia cumingiana	*kelat asam*
Eugenia claviflora	*kelat merah*
Eugenia longiflora	*kelat*
Psidium guajava	*jambu batu*
Pimenta officinalis	*bumbu cengkeh*

*Syn. *Syzygium aromaticum*

6

The Nutmeg Family Trees

OF the commercial nutmeg species, whose seeds are large enough and aromatic enough to be worth selling, four are native to the Moluccas—like the particular *Eugenia* called the clove, and two to New Guinea. The two species from the northern Moluccas have excellent mace, but the most popular and frequently cultivated nutmeg, *Myristica fragrans*, seems to come from, or has always been most grown in, the Banda Islands. This small group of islands lies about 210 km south-east of Ambon in the Banda Sea, between Ceram to the north and Timor to the south. Alfred Russel Wallace, recounting in *The Malay Archipelago* his trip to the Bandas in the 1850s, describes the impressive volcanic cone of Gunung Api rising sheer out of the sea for 700 m, the incredibly clear waters and delicate coral (now no more) of the fine natural harbour formed between the volcano [island] and Banda Island [itself], and the beauty of the nutmeg plantations full of glossy-leaved trees bearing their golden-rosy fruits. At any one time, he notes, a certain proportion of the fruits, being ripe, would have split open at the bottom to reveal the dark brown shell of the nut and the bright red mace which twines around it (see Plate IV). Like all the nutmeg family, the trees of Banda are smallish evergreens with a 'tidy' conical shape; when their fruits dangle so ornamentally from their branches, they really are a very pretty sight.

Nutmegs, the Bandanese say, cannot grow far from the sea; certainly they are rarely found growing at more than 1 000 m above sea-level. They are, however, a pan-tropical family of more than 350 species, 240 of which are native to lowland Malesia and all of whom are, if female, easily recognizable by their fruits and nuts—all variations on the theme of the commercial species in the matter of size and aroma. The flowers also are remarkably similar throughout the family; they are small, pale and inconspicuous, but unusual: they are so simplified that they have 'lost' their petals, and most have only three sepals. Since particularly the female flowers and the fruits are so alike, finding one or another is a good way to identify a tree as a female nutmeg; the males are not so easy to identify and one must look at other characteristics like the shape of the tree, the blue-green colour of the leaves with their many side veins, the tell-tale sap (discussed later), and the hairiness of the young twigs; like some species of *Artocarpus* and durian, they are covered with fine brown hairs. In the case of nutmegs, these completely disappear when the twig bark hardens.

The fact that male and female flowers are always, or almost always, found on separate trees is a condition called dioecious or 'two-housed'. Trees of both sexes are therefore necessary for the production of nuts, which only the female tree bears, unless one of the rare bisexual trees happens to be present. Some plantation owners like to count on the presence of the odd bisexual tree to provide enough male flowers without their having to bother to plant male trees, but this is bad genetic policy. Different pollen carries different properties such as the ability to resist disease, and if the parent stock is not varied from time to time, most easily in the case of nutmegs by planting new male trees, then the whole plantation will become more homogeneous, and consequently more vulnerable: every tree will succumb to the same disease if attacked because every tree will be practically the 'same' tree, genetically speaking.

To keep variability and their genetic ability to respond to adversity, plants must be able to cross-pollinate. Although nature usually arranges that self-pollination is impossible, either by the position of the male and female parts, in the case of bisexual flowers, or by staggering the maturation times of the male and female flowers in monoecious plants, dioeciousness could be regarded as a foolproof device to ensure cross-pollination, and it is particularly common among rain forest plants. Precisely because there are so many different species in any one area of undisturbed rain forest, the distance between individuals of one species is usually large; a plant able to pollinate itself, therefore, would probably do so and inbreeding would result. Dioecious trees do not have this option and must have the assistance of insects or animals in order to reproduce themselves.

The bark of nutmeg trees is characteristically thin, and when cut or slashed oozes a watery pink sap which, on drying, darkens to a deep browny red exactly the same colour as dried blood. A pathside tree slashed by an idle passer-by looks as though some wounded animal had bled on it and accounts for the Malay name of *pendarah* (*darah* = blood) for this family. *Pendarah* also has connotations of 'strong' and 'magical'; in Indonesia, the word is often applied to a *kris* said to have magical properties. The blood-like sap and the aromatic nature of the seeds both encourage a belief in the nutmeg's medicinal uses, e.g., one of the New Guinea species, *Myristica argentea*, is used as a cure for headaches, and carrying a nutmeg on one's person is said to be good for muscular aches and pains and rheumatism. Nutmeg oil, and nutmegs are 50 per cent weight for weight oil, is also both pleasant smelling and an excellent illuminant, but the nuts are more commonly used as medicine than any other way in their place of origin.

These thin-barked, small-flowered and usually small-leaved trees would be considered quite 'advanced' from an evolutionary point of view (see Chapter 8 for a discussion of the 'modern' versus the 'primitive' tree) were it not for the two very primitive character-istics they still retain. The first is the *aril* or edible coating of the seed, sold as mace, or *bunga pala*, in the commercial species, and usually a bright eye-catching orange or red; the second is that the fruit is dehiscent, that is, it splits open when ripe, displaying the nut with its red aril, and by this advertising its ripeness. Dehiscent fruits which split to emit either a strong, ripe smell like the durian, or reveal red arils like the nutmegs, are typical of plants which need the services of birds and other animals to disperse their seeds. In the Moluccas, the 'dispersal agents' (aside from man) are mostly the nutmeg pigeons, *Ducula perspicillata*, *D. bicolor* and *D. concinna*, which feed on the mace but drop the hard-shelled nut on the ground; other animals must perform the same service elsewhere. Plant–animal associations are characteristic of the rain forest environment and are of course essential to the survival of both the plant and the animal. Most modern fruits have evolved away from this depen-dence by a progressive loss of aril and a drying up of the seed cover into a pod or a capsule, or, via indehiscence (not splitting open) towards the succulence of berries, whose small seeds are dispersed more easily. Needing assistance in pollination is not so unusual, but having to attract the larger fauna to disperse fruit and seed is a 'primitive' or little-evolved characteristic of the nutmegs, and it confines this family to the rain forest environment just as surely as it does the wild durians.

*   *   *

Now for the rest of the spice story and the end of the Dutch monopoly. Spices were, of course, the principal 'wealth of the Indies' over which the European nations had been squabbling, and the aim of every maritime nation had been at the very least to by-pass all the middlemen. Obviously, one of the Dutch East India Company's first concerns was control over both the sources of and the trade in cloves and nutmegs. By 1650 they had more or less established their terrestrial monopoly, but it was not until 1669, when they finally sacked Makassar in south Celebes (Sulawesi) that the local trade routes were secured. For nearly ten centuries, the Bugis people, the piratical Phoenicians of the archipelago, had dominated the Straits of Makassar and the Java and Banda Seas from their famous port. The Bugis are still, like the Bajo (or Bajau) people, the sea-traders of Indonesia, and their big craft still crowd the docks of Jakarta's old harbour. But with the arrival of the European ships and their superior fire-power, particularly the broadside so devastating to smaller craft, and with the destruction of their principal stronghold, the days of Bugis greatness were over.

By the mid-seventeenth century, clove and nutmeg growing had spread to other islands in the Moluccas, but the Dutch believed that only decreased production would sustain the high prices, and proceeded to enforce the destruction of all plantations except those on islands over which they had absolute control, selecting Amboina (Ambon) for cloves and the Bandas for the nutmegs. Since the Bandas were the original home of the nutmegs, this was a happy and logical choice, but cloves, the local people say, 'need clouds to grow', which the flat hills of Amboina did not provide; clove production dwindled steadily during Dutch rule.

The Dutch could not hope to control the spice trade forever, but by these methods they succeeded for the next eighty years. They were eventually circumvented by an amiable-looking and enterprising Frenchman with one arm, a keen agriculturalist and amateur botanist, whose great ambition in life had become to make the French possessions in the western Indian Ocean into a second, and rival, Indies. For one so interested in spices, his name, suitably enough, was M. Pierre Poivre.

Poivre's first venture in life was to go to China as a missionary at the age of 20. On his way back to France four years later, his ship was attacked by the English and during the fight, he lost an arm. He was then taken to Java as a prisoner. During his enforced stay, he decided that the clerical life was not for him, and that henceforth he would devote himself to the study of the natural world, whose many facets he admired in Java. As he recovered, he used his time to obtain knowledge of, among a host of other interesting tropical phenomena, the origin and cultivation of cloves and nutmegs. Once free, he joined the French East India Company, and from Île de France (now Mauritius) he set sail once again for the Indies, with the express commission from the Company to obtain live seedlings of the precious clove and nutmeg trees. He seems to have gone instead to the Philippines, where he managed to procure some fresh nutmegs through Chinese traders. It was not until his second mission, some twenty years later, that he went to Timor, where he made an agreement with the Portuguese governor to deliver to him as many 'spice plants' as he desired, under the very noses of the Dutch. These plants he established in the French islands of Mauritius and Réunion, and later in the Seychelles and Madagascar. His dream of rival spice islands was never quite realized, but the monopoly was broken forever.

7

The Nephelium Fruits or 'Little Clouds': Rambutan and Kapulasan

THE rambutan, *kapulasan*, *mata kucing* and *lici* (litchi) are all domesticated species of the genus *Nephelium*. The name is poetic: in Greek, *nephelion* means 'little cloud', and it is a good if vague description of the delicious pulp surrounding the seeds of these fruits. Usually milky-translucent in appearance, it has a pleasant acid-sweet taste and a juicy and succulent consistency, neither hard nor soft, which makes the fruit very refreshing on a hot day, satisfying but never cloying. The litchi is perhaps the best, but it is possible to eat a lot of rambutans and never tire of them, and some think the *kapulasan* to be even better.

With the exception of the litchi, which is a native of southern China, all the known *Nephelium* species seem to have originated in the forests of the lands once joined by the Sunda Shelf (Malaya, Borneo, Java and Sumatra). Again with the exception of the litchi, none of these trees will grow at altitudes much above 300 m or 400 m, or in places with a rainfall of less than 3 000 mm a year, which effectively confines them to a lowland rain forest environment. Their wild relatives are still to be found growing wherever sufficient rain forest remains; out of the 35 known species, 25 are found in the forests of Borneo.

The wild and domesticated species resemble each other closely, and *Nephelium* trees even when not in fruit are easy to recognize. Varying between 10 and 20 m in height, they have a rounded shape, untidy-looking crowns and dark green leaves which are compound and pinnate, that is, each 'leaf' is really a series of leaflets, usually two to four pairs on a stalk. Like most members of the *Sapindaceae* family to which they belong, the leaves are always even in number since there is no single leaflet at the top. The leaflets are also rather narrow in relation to their length, and tend to curl upwards along their edges, and the young twigs are distinctly grooved: these are all good ways to identify a *Nephelium* tree when it is not in fruit.

The rambutan is one of the commonest South-East Asian village trees. During the fruiting season, normally from November to February, the ripe fruit is very eye-catching as it hangs in bunches among its dark leaves, or is offered for sale in great baskets-full by the road side. The rich, strawberry red is the most usual, but there is also a yellow, said to be a Thai variety; people often plant one of each for the handsome colour combination they make, and add a dark red *kapulasan* for contrast. All of the rambutan fruits have the same distinctively 'hairy' appearance, however, even the wild ones—the name means 'hairy' in Malay and Indonesian—and the nuts of all varieties are also useful in that they contain a fine oleic acid, or oil, which constitutes on the average 37 per cent of their dry weight. At room temperature this oil is white and hard and is therefore good for making soap or candles, but it is little valued in these days of cheap substitutes.

Kapulasan fruits, by contrast, have blunt knobs instead of hairy 'processes' and a yellower, sweeter flesh which generally separates more easily from the nut than it does in the rambutans; there are even some seedless varieties known. The nuts also contain oil and, if roasted, would make a substitute for cocoa, but of course a seedless fruit would have the most commercial value. Fresh *kapulasan* fruits are the more frequently exported of the two: without the hair-like spines, it is more economical to pack and ship them; peeled rambutans are tinned in Thailand and Malaysia.

The chief differences between the species, and to some extent within each species, are the colour and appearance of the outer skin of the fruit, the thickness and colour of the pulp, its sweetness and the ease with which it can be separated from the seed when it is ripe. Of rambutan fruits alone there are said to be twenty-two varieties in cultivation in Indonesia. Most of them have been produced more by luck than intention, since the trees cross-fertilize indiscriminately. *Nephelium* trees have a very odd sort of sex distribution, neither wholly bisexual nor monoecious or dioecious[1] but a mixture; the fruiting trees have bisexual flowers while others, which do not set fruit, have only male flowers. Because in nature self-fertilization is usually prevented by some mechanical means (the flowers' structure or delayed maturation times), *Nephelium* trees are

[1]These terms are explained in the chapter on nutmegs and in the Glossary.

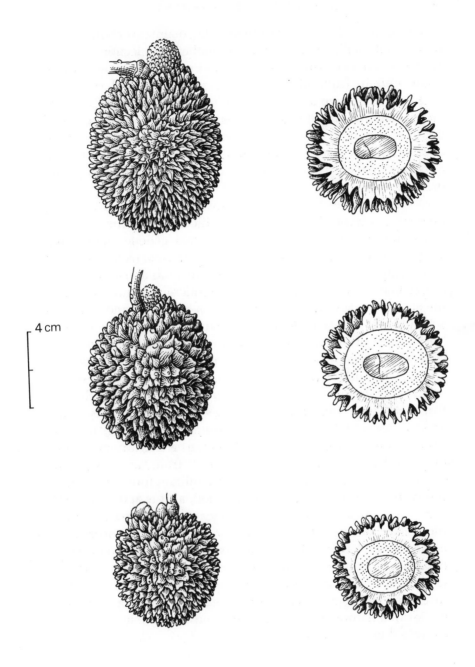

Fig. 12 *Nephelium mutabile (kapulasan)* fruit

in fact compelled to cross-pollinate. This maintains good genetic variety and vitality, but makes the deliberate selection of good 'races' or varieties extremely difficult, as many fruit-growers have found out. In fruits of considerable nutritional and economic importance, the size of the fruit and how easy it is to eat can affect the diet and income of large numbers of people. Fortunately with the *Nephelium* species, each species can be budded onto the stock of any of its relatives and the cuttings of known varieties are successful duplicates of their parent stock; it is breeding for improved quality that is difficult.

Nephelium species have certainly proved susceptible to selection even on an informal level despite the difficulties, and the genus is a worthwhile target for the more controlled genetic engineering attempts now being made. Breeding for certain characteristics, and then establishing the plants which have them as a new species by inbreeding, is a familiar process in scientific agriculture. Basically, the scientist is speeding up a process of mutation and selection which might happen anyway by manipulating the genes, the 'messengers of heredity' which pass on characteristics from one generation to the next, in order to improve the nutritional and commercial value of a plant (or an animal). When he is successful, a more hardy strain is the result. But it has often happened in the past that a new variety or species combines its better qualities with a greater susceptibility to disease or some other failing, and the scientist must return to the wild plants of the genus for new genetic material. For the rambutan and *kapulasan* fruits, as for so many other cultivated tropical fruits, the lowland rain forest is the great reservoir of this material: the 'gene pool', as it is aptly called. Of course it is also the gene pool for species as yet unknown, some of which may become extinct before they can even be discovered and named by the hard-working botanists. It is for this reason that the botanists, even those who work only with cultivated species, are just as anxious as the conservationists to try to preserve enough self-sustaining and representative undisturbed lowland rain forest in the form of nature reserves and parks, now, before it has all been converted to other uses.

8
The Durians and the Durian Theory

THE spiky fruit of the durian, *Durio zibethinus*, is the delicacy of the Sunda region. The rich, creamy pulp is prized beyond all other foods—prized, that is, by those who are not adversely overcome by its smell. Wallace declared that to eat a durian was well worth a voyage to the East for this reason alone! Although he added, as becomes a fair-minded scientist, that those who dislike it compare it to custard passed through a sewer. Its smell is so strong that local airline companies forbid passengers to carry the fruit, and you often see, in hotel bedrooms, a terse note from the management: 'No durians or outside guests of opposite sex allowed.'

Obviously, it is not possible to remain indifferent to the durian, and no one is. Its local popularity makes it hard to understand why it is usually only *D. zibethinus*, a species only found in cultivation, which is grown and so rarely any of the other twenty-seven species of the Malesian rain forests—in spite of the close resemblance of the wild and cultivated species, and in spite of the fact that wild durians are much appreciated by the jungle-dwelling peoples and animals. Botanists also are no strangers to their merits. Dr A. J. Kostermans mentions the excellent qualities of both *D. oxleyanus* and *D. kutejensis* in his beautifully illustrated monograph on the genus, published in the Bogor Herbarium's journal (*Reinwardtia*) in 1958. *D. oxleyanus* is the favourite food of Sumatran orang-utans, and I prefer it myself since it is just as sweet but not a tenth as odorous as *D. zibethinus*. Wild durians are found throughout Sumatra, Malaysia and Borneo, but Borneo, with the most endemic species—fourteen out of the nineteen found there—is probably their land of origin; it is obviously in Borneo, therefore, that it is most worthwhile to look for new species to bring into cultivation. Unfortunately durians will only grow between sea-level and 700 m, and it is just these lowland rain forests that are subject to the heaviest exploitation for their

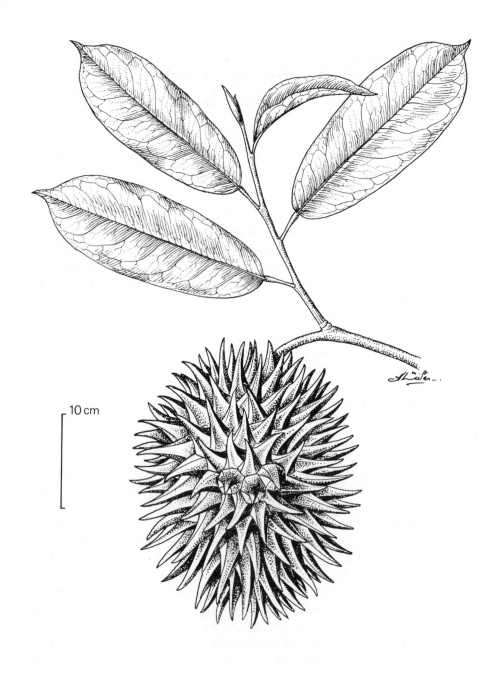

10 cm

Fig. 13   *Durio oxleyanus*, a wild durian

timber, making *Durio* the most endangered genus among the popular fruit trees of this part of the world.

Durian trees are reasonably easy to recognize in the forest as well as in cultivation. They are stately trees, often very tall, with large straight trunks and thick but attractive bark in tones of pale russet and grey, and their rather narrow crown looks very dark, even gloomy, because of the exceptionally deep olive green of the leaves. These leaves can be up to 20 cm long, grow alternately on the twigs and are always silvery or copper-coloured underneath, an effect of the small scales that cover their under-surface. When a tree is in bloom, its large white flowers are borne in clusters mostly on the thicker branches or the trunk (=cauliflory), and when they are open, they smell strongly of sour milk. White flowers which smell thus are nearly always fertilized by bats, and although durian flowers are visited, when they open in the late afternoons, by various insects (honey bees, flies, beetles, lace-wing butterflies), the pollen grains have been found to form a large part of the diet of only one species of bat, *Eonycteris spelaea*, a long-tongued fruit bat. Other bats also visit the flowers but only seem to chew the flowers destructively (according to E. Soepadmo and B. K. Eow), and the likely conclusion is that in Peninsular Malaysia at least, this particular bat is the pollinating agent. Fruit bats of the *Eonycteris* genus are the smallest of all fruit bats and have mouths and tongues specially adapted for feeding on nectar and pollen from flowers which, in their turn, are also specialized, the most distinguishing characteristic in their case being their long, projecting stamens. *E. spelaea* fortunately has a wide range in South-East Asia or some of us would have to live without durians!

On the other hand, a bat feeding on durian pollen must indeed have to live without it a good part of the year. What keeps it alive during all those months in which the durians are not in flower? Apparently, it also visits *Bombax* and *Ceiba* (kapok) trees, which are in the same botanical family as the durians, as well as *Artocarpus*, *Parkia* and *Duabanga*; obviously, the animal pollinator of annually-flowering trees must depend for its survival on a variety of tree species, and it also follows that these tree species must continue to exist (in this case as food sources) if durians, whether cultivated or wild, are to set fruit.

The durian fruit, as every South-East Asian knows, is a large, round or ovoid fruit weighing between 2 and 4 kg and covered with thick-set and very sharp spikes. This heavy armour is presumably intended by nature to protect the big seeds inside until they are fully developed and thus avoid the waste of genetic material; in spite of it, orang-utans, who are fond of unripe durian, have devised ways of opening the capsule, but most wild animals have to wait until it splits naturally, displaying the edible pulp and exuding in full strength the heavy pungent odour the fruit is famous for. At this time also the pulp is thought to have attained its ultimate perfection in taste and consistency. Attracted by the gradually increasing odour, animals, particularly elephants, will gather under a tree with ripening fruit, and jungle-dwellers who know the location of wild durians will usually clear the undergrowth and camp there themselves several days in advance to keep the animals away. The time of ripening is critical; a few hours after the fruit has opened, rapid chemical changes take place which completely alter its consistency and taste.

Under natural conditions, that is, in the forest, the pulpy arils are eaten up at once by the waiting animals. This releases the seeds for immediate germination, which takes place within about eighty days. Like *Artocarpus* seeds, durian seeds have no dormancy period and therefore cannot germinate at all if the process is not started at once—they rot instead. Since they are large, they are not carried far from the tree, and young trees are therefore usually found growing near the parent tree; this limited type of distribution ensures the plants good growing conditions or at least suitable ones, and is also good for later cross-pollination. Durian flowers are bisexual, but self-pollination is prevented in this case by the fact that the anthers or pollen-producing sacs are positioned below the ovule. A pollinating agent is therefore needed to transfer the one to the other, and the presence of other durians reasonably near means that genetic variety is easily maintained by the bats as they pass from one flower to another.

Durians, even the domesticated species, do not seem in much danger of over-refinement of their genetic material: the quality of durian fruit is well-known to be very variable. Some of the trees in cultivation produce such delicious fruit that they are known by name and location and their seeds command a high price from the ever-hopeful purchaser, even though the seeds hardly ever 'breed true' to produce another tree exactly like the parent. This fault, often found among fruit trees, is usually avoided by budding or grafting the known stock on new stems; but grafting rarely 'takes' between durians and it has not been a solution for this genus. It is possible that if more than one species were regularly cultivated, some more stable hybrid could be developed. But, and this is where the danger to the genus lies, with so much of their natural habitat felled every year many of the wild species are being so reduced in number that, inevitably, they will soon become extinct.

THE DURIAN THEORY

In Chapter 6, the durian was mentioned as one example of a 'primitive' fruit. Corner, the lively botanist who worked in Malaya before, during and after World War II, decided after some thought that it was the outstanding example. In 1949 he introduced into botanical literature what he called 'The Durian Theory'. This was his theory of the evolution of the modern fruiting tree; he went on developing the idea in a series of publications over the next few years.

According to generally accepted evolutionary theory, complex forms of life have developed from simpler forms by a gradual process of change induced by 'natural selection', the phrase Charles Darwin made so famous in 1859 in his book, *The Origin of Species*. Darwin and his friend Wallace had both observed that all species vary slightly between one individual and another, and that in competition with each other and with different species, some individuals possess characteristics which enable them to survive existing conditions better than others do. Naturally, those who survive pass on their genes to a new generation, and cumulatively this results in changes in the species towards a perfect adaptation to its environment. On the other hand, the environment also changes over long periods of time and to cope with this, continued genetic response to change tends to produce changes in the genes—mutations—which if successful eventually become new forms and new species.

The forms of modern trees have, according to Corner's theory, evolved in all their present variety from an ancestral type of thick-trunked, relatively short tree bearing large and often complex leaves on thickish twigs. A thick stem grows slowly, does not branch readily and has little flexibility in the manner of its branching. Such a tree will also never reach any great height, especially if it is dependent for photosynthesis on large but costly-to-produce complex leaves. The papaya tree still has several of these features. Therefore 'progress', or the evolutionary trend in trees, has been

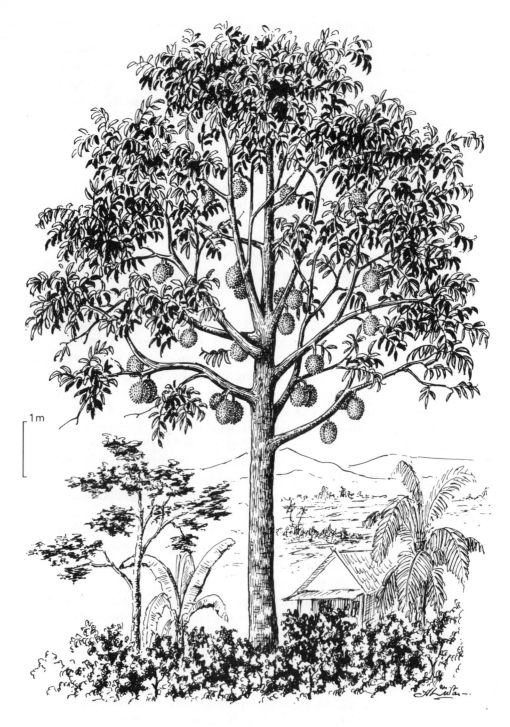

1 m

Fig. 14   *Durio zibethinus* (durian), a tree in fruit

towards taller trees with thin bark, able to gain height quickly over competitors by fast upward growth while still slim. Only then do their trunks thicken at the base, providing support, while their crowns divide into competition-shading branches whose twigs bear easily produced and therefore easily renewable small, simple leaves. Leaves are the food factory of a plant; height, access to the sun and their production/replacement at no great cost to the plant all have tremendous survival value.

As for the fruit, Corner reasoned that the primitive fruit type was that of a large, spiky capsule which only splits open when ripe—plants, particularly those producing big heavy fruits, must be economical and prevent waste—but then offered to its animal dispersers tasty, rich, strong-smelling seed coverings, often also 'advertised' by bright, eye-catching colour. For such large seeds, animal distribution is obviously necessary, and fairly large animals at that; this the rain forest can provide, and those trees which remain are by and large those which need such animal services the most. Progress, in the modern fruit, has been a development away from this dependence: by decreasing size, by a progressive drying up of the aril and the capsule, towards ever more portable, even wind-blown, seeds, or towards the succulence of berries. It is hard to imagine greater opposites than the heavily-armed durian and the naked, succulent berry. As Corner suggested, in tropical forests where there are still examples to be found of every stage in the development of the modern tree, the durian remains behind, an evolutionary signpost.

9
The Artocarpus Trees:
Breadfruit and Jackfruit

Two species of the *Artocarpus* genus, the jackfruit or *nangka* and the breadfruit, *kluwih* or *sukun*, are cultivated throughout the tropics. Because they bear large fruits throughout the year, they are an important source of food wherever they grow and have been much introduced for this reason. Where did they come from originally, these useful trees? Travellers were reporting the jackfruit in Central Asia as early as 300 BC, and a number of its close relatives occur, apparently indigenously, throughout tropical Asia. The jackfruit itself is probably not Malesian, however, but was introduced into Malaya too long ago for any record to remain. The similar *cempedak* is indigenous to Sumatra, Borneo, Sulawesi, West Irian and the wetter parts of Papua New Guinea, but is not found in Central and East Java, the Lesser Sundas or eastern New Guinea, all of which places have a marked dry season. The breadfruit and its close relatives seem most adapted to a mildly seasonal 'monsoon' climate, providing the dry season is not severely dry. By the end of the seventeenth century, breadfruit cultivation was so widespread in the Pacific islands that it formed a staple of the Polynesian diet, and was much remarked on by travellers, particularly by those who accompanied Captain Cook on his first voyage around the world (1768–71). By 1787, Captain Bligh, the notorious Bligh of the ship *Bounty*, had been sent to Tahiti to take on board a thousand breadfruit seedlings for transport to the British West Indian colonies; in his anxiety to get them there quickly, he would not allow the sea-weary sailors long enough ashore and once at sea drove the ship hard, which involved long hours and many a weary sail change to get the best out of the wind; which triggered the famous mutiny that followed.

Rain forest origins for most if not all fifty known species of *Artocarpus* are likely, however. Most *Artocarpus* trees grow best in hot,

Fig. 15 Breadfruit on the *Bounty*, an engraving by Robert Dodd; 'Captain Bligh and party cast adrift by the mutineers from *HMS Bounty*', c. 1790

wet climates and most of them retain decidedly 'primitive' characteristics, such as large perishable seeds, relatively large (and very
large, in the case of the jackfruit) fruit with edible arils and a
tendency towards cauliflory: all of these are qualities linked to a
seed dispersal system dependent on animals. Cauliflorous trees
bear their flowers and seeds on their trunks or thicker branches
where it will be most visible and accessible to the larger forest
fauna.

Cauliflory is particularly noticeable in the case of the jackfruit.
The fruits develop from buds on the trunk and lower branches in
this case, one suspects, not only so they will attract animal
dispersers of their seeds, but also because such fruits must need
gross supplies of food and water in order to form. Some of them
weigh up to 30 kg and have to be carried to market on a pole
between two men, trussed like a pig. Naturally they also take quite a
long time to mature and when nearly ripe, they are lovingly looked
after. A sight which intrigues many visitors to Java is the number of
trees along the roadside which seem to have large, obviously full,
plastic, paper or cloth bags tied to their trunks. Are some strange
and secret goods being offered for sale? Or do the trees stand at
recognized meeting places where footsore travellers can hang up
their bundles for a while? Not at all. It is only by such careful
wrapping of the fruit that it can be guarded from the nightly
depredations of the giant fruit bats, the *kalong*, or 'flying foxes'.

Nearly all *Artocarpus* trees share a peculiar shape of flower head
which resembles a bur on a stalk and which bears flowers so small
that they have to be dissected under a microscope to tell one part
from another. Flowers of both sexes are found on the same tree and
usually on the same twig (though not in the illustration of the female
jack flowers here). The male flower head is round and smaller;
when mature it is covered with yellow pollen and the tiny stamens
can just be seen if you are close enough. They fall off after flowering;
the female flower head enlarges to become the fruit.

Artocarpus leaves are also distinctive, though there are great
differences in size. The young leaves tend to be larger than the older
ones, markedly lobed or indented and of a bright green colour
with yellow veins, in contrast to the smaller, darker, unlobed or
'entire' leaves of the adult tree. The leaves of the *terap*, *Artocarpus
elasticus*, are a good example of this: some of the sapling leaves are
almost 2 m long! There is evidence, however, that this is not actually
a question of age but of the distance the leaf is from the ground,

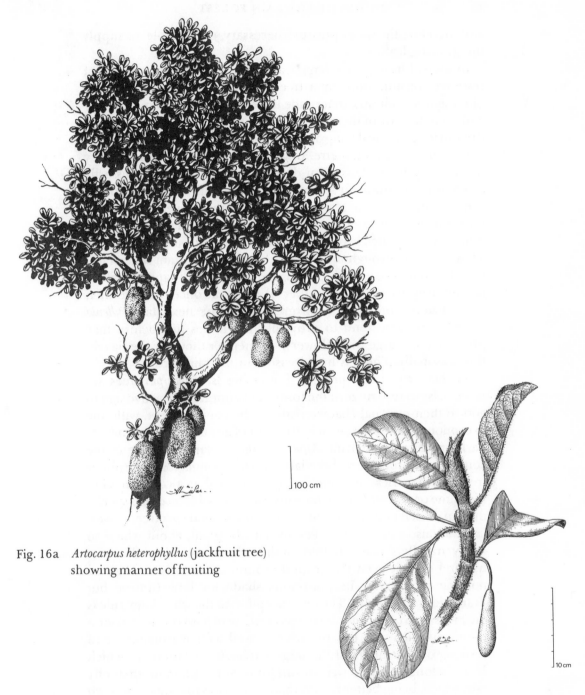

Fig. 16a *Artocarpus heterophyllus* (jackfruit tree)
showing manner of fruiting

100 cm

10 cm

Fig. 16b Jackfruit twig with flowers,
showing their peculiar shape

and therefore the water pressure necessary, or available, to supply the growing leaf.

In spite of having such large leaves, which are costly in terms of resource-consumption for a tree to replace, *Artocarpus* trees are often semi-deciduous, that is, the leaves will change colour slightly and then drop from the tree at regular intervals. Crunching over a dry 2 m long *terap* leaf lying on the jungle floor is not an experience I have actually had, but metre-long ones are very common and they make noisy litter. This semi-deciduous habit is not necessarily linked with monsoon, or seasonal, climate distribution; some of the species adapted to a seasonal climate remain evergreen, while others in ever-wet conditions put out 'flushes' of new leaves. True rain forest trees shed leaves as well, of course, but usually shed them a few at a time throughout the year, being in a constant state of leaf die-off and renewal. Only some figs will occasionally appear naked, having dropped a great number of leaves at once, but are then immediately clothed in fresh green ones the next day. *Dillenia* (*simpoh*) trees also tend to drop all their leaves just before they produce their large waxy flowers, but neither they nor the figs do this seasonally. There is little 'season' in the rain forest.

By their leaves and peculiar flower heads, *Artocarpus* trees are rather easy to recognize in the forest, but another distinctive sign to add to their external characteristics is the copious latex, with one exception a white latex, which drips out of any cut surface, whether bark or leaf stem or fruit. *Artocarpus elastica* sap has some of the commercial properties of the latex of *Hevea brasiliensis* (the rubber tree of S. America); it is sticky and often used as a component of bird lime or dart and arrow poisons. In some wild *Artocarpus* species the sap is itself poisonous—like that of the related *upas* tree, *Antiaris toxicaria* (also called *ipoh* in Peninsular Malaysia), about which so many 'travellers' tales' are told: of the swift death which follows its use and of how even the ground around its roots is so blighted nothing can survive its poisonous shade for long (untrue but dramatic in the telling). The one exception to the white latex rule is the greasy sap of the *miku, Artocarpus lowii*, which is known to make a useful ointment for lips or sores as well as a reasonably good cooking oil. Members of the large *Euphorbiaceae* family, to which *Hevea* belongs, are also well known for their latex and for their oily saps which, being better known than *Artocarpus* saps, are put to such commercial uses as illuminating oils, soaps and 'tung oil', an important ingredient of paints and varnishes. So far, the only

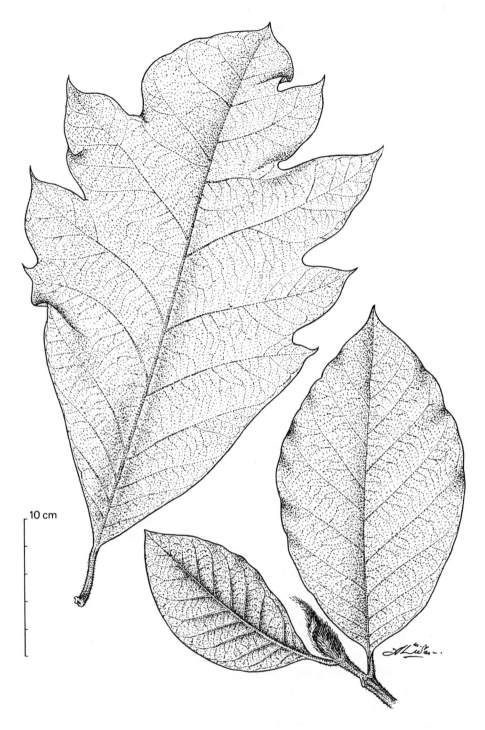

Fig. 17 *Artocarpus elasticus* (*terap*), sapling and adult leaves

common commercial use of any *Artocarpus* has been, aside from selling the fruit, the jackfruit's fine timber: yellow when new but darkening to a mahogany colour on aging and excellent for furniture since it resists insect attack.

Artocarpus fruits come in many sizes, shapes and coverings. Some wild species have velvety skins and pink flesh: *miku* fruits are small, prickly and yellow, the jackfruit and *cempedak* a prickly olive green, and the breadfruits are variously nubbly green or smooth, marked only with the pattern of small hexagons characteristic of the genus. Like the pineapple or the *pandan*, the fruits are not formed from a single fertilized flower but from syncarps which are compacted female flowers joined together on a common base; the hexagons are the scars of their former separate floral existence.

Only about half the species of this genus are cauliflorous like the jackfruit; the others, like the breadfruits, are borne 'normally' at the ends of branches. You can always tell the breadfruits from other *Artocarpus* trees by the leaves: only breadfruit leaves retain their indented or lobed shape when adult and it is mainly because of these leaves that the adult, fruiting tree is such a splendid sight. Its few but sturdy branches arch overhead and the big shiny leaves glint in the sunlight, each one well-displayed by their spiral growth habit; in the centre of each 'spiral' of leaves, borne on a short stout twig, is the yellow-green fruit. Of the two types of breadfruit common in Indonesia, the *kluwih*, which has seeds, is usually found in the areas with highest rainfall, while the *sukun*, the seedless variety, is more likely to be found in drier areas and more closely resembles the breadfruits of the Pacific and Indian Ocean islands. *Sukun* fruits are very tender and squash easily when they fall to the ground, yet will not ripen off the tree to any great extent; to gather them, one person must dislodge them while another waits below to break their fall by the dexterous flip of a gunny sack at exactly the right moment.

Except for the *sukun*, all species of *Artocarpus* have large seeds; though they are not as large as the durian's, they are just as edible and are considered almost as good as the fruit itself in taste and food value. Correspondingly, again like durian seeds, they will spoil quickly unless they germinate or unless they are 'killed' by roasting or baking them. Like most rain forest seeds, they have no dormancy period and cannot be desiccated or otherwise preserved for planting later. But unlike the durian trees, the *Artocarpus* species take well from cuttings, root suckers or layerings; their seeds are hardly necessary for their cultivation, and this accounts for the ease with which they have spread around the tropical world. The almost seedless varieties of domesticated breadfruit, therefore, are usually the most popular because the seeds, though tasty, are redundant and their absence means a greater proportion of the starchy flesh. It is, of course, the texture and use of the flesh (or aril) when cooked that gave breadfruit its common name and science the name of the genus: in Greek, *artos* = bread and *karpos* = fruit. In many of the countries in which it grows it literally is 'the staff of life', though this line of reasoning would not appeal to those for whom rice is the staple food.

Artocarpus and durian trees obviously have several similar features besides their seeds: durians are not the only large prickly fruits of the rain forest nor the only ones with rich tasting, strong-smelling arils or cauliflorous habits. Corner thought the jackfruit an excellent example of a sideline of evolutionary development. Jackfruits illustrate the way in which a tree which has evolved the modern twig and slender trunk of an 'advanced' type of tree can still retain the habit of bearing large, even huge, fruits with succulent arils. Either, as he explains in *The Life of Plants*, the 'fruit became small and kept its position on leafy twigs . . . or it remained large and massive, developing only from dormant buds on the branches and trunk where there would be sufficient food for its supply'. Figs and other cauliflorous genera or species that have not retained either the arils or the size do not come into this category. Corner also remarks that in their transition from relatively thick-trunked saplings with few branches and large, much-indented leaves, which is typical *Artocarpus* 'architecture' when young, to tall relatively slender adults with more branches and simple, smaller leaves, the jackfruit trees provide a living parallel of the stages through which the modern tree, evolving over millions of years, has passed.

10
The Agathis

IN 1653, a young German with large, penetrating eyes and an even larger nose, very much the beaky, bony European, landed in Java. He had newly enlisted as a soldier in the service of the V.O.C., the Dutch East India Company. That same year, he was sent to the station of Amboina (Ambon) in the Moluccas; four years later, he was appointed Amboina's Second Merchant. But Georg Everhard Rumpf did not acquire fame as a soldier or a merchant. Herr Rumpf, or Rumphius, to use the Latinized form of his name as was a popular fashion of the time, collected: birds and butterflies, beetles and insects, animals and plants. Any aspect of the natural world fascinated him, but particularly, one supposes, the plants, for he created single-handed the famous Herbarium Amboinense, and was the first European to describe scientifically over a thousand tropical plants never seen before or heard of by any botanist. For the rest of his life he continued to collect, describe, compare and preserve. Twice his collections and drawings were destroyed by fire, but he managed to restore or replace them even after blindness overtook him (in 1670), helped in his last years by his devoted wife and assistant, and his definition and description of some of the genera he first named has even now hardly been improved upon.

Under the Latinized Malay name of *Dammar*, one of the genera he was the first to describe was *Agathis*, the best known of the few Malesian conifers adapted to an ever-wet climate. They have only one relative, the *Araucaria*, which genus includes the monkey-puzzle tree so popular in English suburban gardens; most *Araucaria* trees are found in New Guinea. The two other conifers growing in the rain forest are *Dacrydium* and *Podocarpus*; like *Agathis*, they are found mostly in the mountains or on poor soils in the lowlands, on what one might call second-rate sites.

Conifers, incidentally, are seed plants but not flowering plants. They are gymnosperms, from the Greek for 'naked seed', and along with two other families of naked-seeded plants, they are the surviving remnants of what was once the dominant plant form of the world. Gradually the gymnosperms were displaced by the *angiosperms*, or flowering plants, whose seeds are enclosed in an ovary. The angiosperms became the more common type of plant because they were more diverse in habit, more adaptable and therefore more successful ecologically in competition with the gymnosperms. Now they dominate all land vegetation, while the gymnosperms are restricted to areas where for special reasons they have an advantage over their angiosperm rivals.

Agathis trees are found today from the Malay Peninsula to the New Hebrides, Fiji, New Caledonia, south Queensland and in New Zealand. But they are not native to Java (most Javan *Agathis* originally came from Ambon), nor to the Lesser Sundas, the Kai and Aru Islands and Irian Jaya. They are, or were, abundant in Kalimantan, however, particularly in Central Kalimantan where there were once some of the densest *Agathis* forests in the world, and there are still scattered large stands in Celebes and smaller ones in Papua New Guinea and New Caledonia. The southernmost *Agathis* population is the remnant which survives of the once-great forests of the northern peninsula of North Island, New Zealand. Everywhere they grow, or grew, they have been and still are a natural target for exploitation.

For the *Agathis* is, even more than the dipterocarp, the timber man's dream of a tree. Typically conifer in shape, at least until they are quite old, *Agathis* are tall, straight, single-stemmed trees with only slightly tapering boles. The wood is fine-grained and dense, yielding an extraordinary 2 300 tonnes of timber a hectare, which consistently fetches higher prices than that of the popular dipterocarps. They also delight commercial foresters by growing by preference in thick stands and by being exceptionally easy to recognize: from the air, by the greyish tinge of their leaves which contrasts with the general blanket of green, and their usual location on ridge sites; from the ground, by the way their grey brown bark scales off the trunk in flat, disc-like pieces to leave a pattern of irregular concentric scars, and by their smooth leathery leaves with almost parallel veins. They are, in fact, exceptionally beautiful trees, especially when the angle of the sun highlights the pattern of the bark, and in Java they are often planted in avenues for ornamental purposes, as they have been along one side of and in the Botanic Garden in Bogor.

The most common species in Indonesia is *Agathis dammara*, the

Fig. 18 *Agathis dammara*

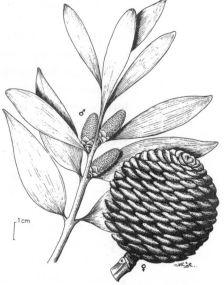

Fig. 19 *Agathis*, male and female cones

Ambon species Rumphius first identified and named after the valuable *damar* or resin which can be tapped from the trunk, and which is made into the commercial product known as Manila copal, or copal varnish. Throughout most of Indonesia, resin-tapping is an important source of income to local people, who usually operate on a free-lance basis. Other species of *Agathis* yield similar resins: when the early European settlers came to New Zealand, they used to fell or burn the *Agathis*—the *kauri* trees—then so plentiful, thus clearing the ground the better to 'mine' the huge lumps of fossilized gum which had settled into the earth around their roots. Some pieces weighed up to 50 kg! As in Indonesia, they also hacked and cut the remaining trees to make them bleed more gum and resin; nearly all of the big trees still left standing have been seriously weakened in this way. As for the timber, it was used for everything: the capital city of Wellington was virtually built of *Agathis* planks. Now that the great *kauri* forests are no more and the landscape in the Northlands changed irrevocably by the European settler, there is much regret and nostalgia among New Zealanders for the huge trees which the Maori people, who were there first, treated with so much greater respect and even reverence.

Since in these days, wild populations of *Agathis* are most often found in mountainous country, it would be easy to assume that they have a preference for hilltops, ridges or steep inclines. This seems to be genuinely true of the Malayan *Agathis*, since there is only one record of a tree from the plains, but it is certainly not true throughout the *Agathis*' range; the hills are often the only places where they have survived. At low altitudes, however, they showed a preference for very poor soils: in Kalimantan, for instance, they grew mostly on the white sand ridges which run through the peat forests of the whole of central Kalimantan. White sand ridges form a sort of 'heath' where an acidic surface layer of bleached silicious sand well-leached of iron oxides forms over a harder, impermeable layer. These soils occur where rainfall persistently exceeds evaporation, but are not typical of rain forest soils in general, which are usually yellowish or rust-coloured from their iron oxide content, though still acidic and leached of their minerals. It is interesting that the former type, or 'podzol' as it is called, suits *Agathis*. Or, looked at another way, the thick growth of *Agathis* in an area may contribute towards the creation of this soil type.

In their ability to make use of poor soils, *Agathis* might be said to have a sort of 'pioneer' quality. They are not true pioneers, which are the trees or plants that first colonize a site like the hardy native pines of Sumatra (see Chapter 15), but they do produce valuable timber from extremely unpromising soils as well as from steep gradients. Furthermore, they can sometimes invade areas of *mature* secondary bush which will provide their seedlings with the necessary shade but will not compete with them too much at ground level.

Indeed one of the most curious, and troubling, facts about the present-day populations of *Agathis* in the Malesian area is that the only *Agathis* stands which can actually be seen to be regenerating themselves, that is, where there are seedlings and saplings mixed together, are found on these poor sites. On normal forest soils, one never seems to find any *Agathis* with trunks of under 30 cm in diameter, and how they are, or if they are, reproducing themselves is not known.

Because of their soil preferences, their fine straight timber and their natural tendency, much appreciated by loggers, to shed unwanted side branches during their growth, the *Agathis* has some considerable future as a plantation tree once experiments now in progress find practical ways to lengthen the life of the seed. But as a genus, it is still genetically highly variable (the male cones are almost impossible to identify as to species), and it is occasionally subject to die-back or sudden death, which may be caused by a malignant root fungus. These problems can no doubt be solved in time, but if *Agathis* is to become established as a genus of 'plantation trees' it will be doubly important to protect their wild relatives throughout the range of their natural habitat. For such dioecious trees as *Agathis*, it has been estimated that the minimum number of trees in any one 'species-reserve' should be 500 in order to allow for its natural high degree of genetic variation and to ensure a viable population. But most of the *Agathis* forests of Kalimantan have already been logged, and those of Sulawesi are fast disappearing. It is high time to set aside adequate reserves where *Agathis* is still common[1] before Malesian foresters have to mourn their wild *Agathis* as the New Zealanders do, with guided tours to the remaining 'giants of the forest', and books about the bad old days of unlimited exploitation.

[1]A good start has been made in the Lore Lindu Reserve, central Sulawesi.

11

The Rafflesia Flowers

HUGE and often evil-smelling, totally parasitic on the host plant, producing only flowers and only flowering at long intervals, the *Rafflesiaceae* must be the most grotesque yet astonishing creations in the rain forest.

The scientific name honours Sir Stamford Raffles, the founder of Singapore. Raffles afterwards became Lieutenant-Governor of Java, between 1811 and 1815, a brief interlude in Dutch rule while Holland was occupied by the French under Napoleon, and then between 1818 and 1824 the Governor of Bencoolen, the administrative centre for the British East India Company's possessions in western Sumatra. Always keenly interested in the natural sciences, Raffles missed no opportunity to investigate and record whatever he came across on this subject, as well as on the many other subjects he was so well read in: art, politics, religion, local customs, archaeology; his *History of Java* is a justly famous book. Although he was poorly rewarded for his brief but brilliant governorship of Java by being 'exiled', as he put it, to the Company station in Sumatra, this did not deter him from making the best of his time there. Bencoolen was the place where his hopes of extending British rule over the islands gradually faded, and where, one by one, his youngest children died of fever, but that is another sadder, later story. At this time, in 1818, he was still full of enthusiasm, popular in England and well-respected by the Royal Society, first among British scientific establishments. It is possible that if anyone else had forwarded to the Royal Society an account of some reeking, metre-wide 'flower' found growing in a far-off jungle, he might hardly have been believed.

In this year, Raffles had just been able to persuade his friend, Dr Joseph Arnold, a member of the Royal Society, to join him and to undertake an extensive investigation of 'every branch of natural history' in Sumatra. From the scientific point of view, the interior of Sumatra was virtually unexplored; Dr Arnold, Raffles and his wife

all set out with great expectations. They were not disappointed.
One day, when Arnold was exploring a little apart from the others,
one of the servants came running up to him with, as Arnold tells it,
'wonder in his eyes, and said, "Come with me, sir, come! A flower,
very large, beautiful, wonderful!" ' Following him, Arnold suddenly beheld what he called 'the greatest prodigy of the vegetable
world'. Borrowing the servant's *parang*, he cut it off the vine on
which it was growing and then, with his help, carried the flower,
which weighed nearly 7 kg, back to the camp. Quickly Arnold and
Raffles measured it . The five huge petals were a foot (30 cm) long
each, and they estimated that the nectary in the centre would hold
12 pints (6.8 litres) of liquid. To be quite sure of the surface meas-
urements, they 'pinned four large sheets of paper together . . . cutting them to the precise size of the flower'. During this opera-
tion, a strong smell of tainted beef assailed their nostrils, coming
from the apparently putrescent flesh of the flower.

To have stumbled on this plant coincidentally with one of its rare
flowerings, and to have measured it scientifically in front of
'respectable' witnesses: what good fortune! And though they were
able to preserve only a part of it, and though, sadly, Dr Arnold was
himself carried off by fever before he could even return to
Bencoolen, he had had time to write an informal letter to a friend
describing his find. Once back at base, Raffles added his own
drawings and those of Arnold's to the letter, which he had found,
and sent the news to England. It caused a sensation in the European
scientific world. Robert Brown, the great English botanist of the
time, published a memoir on it in the *Transactions of the Linnean
Society* and gave a talk on it to the Royal Society, and also gave it the
scientific name that honours its two discoverers, *Rafflesia arnoldii*.
But some botanists refused altogether to believe that the *Rafflesia*
flowers were plants or, at the very least, believed them to be
botanically 'insane'.

Even now, not a great deal is known about them, and they are still
regarded as extraordinary in every way. In the process of evolution,
they seem to have been able to discard every irrelevant part. There
are no leaves to process carbon dioxide and sunlight into sugars
and oxygen, or roots to obtain water and minerals from the damp
earth from which to manufacture proteins. There is no stem to
support the flowers and to transport water and food to them. All
that is left are some long strands of tissue which live inside the host
plant—always a *Tetrastigma* vine of the *Vitaceae* (grape vine) family, in

much the same way as the *hyphae* or filaments of a fungus penetrate the host on which it lives—and the huge sexual organ, the fleshy flower, displayed every now and again on an overhead vine or, more usually, blossoming on the dark forest floor.

The colour of the flower, a reddish or purplish-brown flecked with white, does suggest, as much as anything, a large piece of fly-blown meat, and much has been made of the smell of carrion or rotten flesh which the decaying flower exudes, and which seems to attract so many flies and insects to it. Since the lifetime of the open flower is short, and since so many flowers in one locality seem to be of one sex only, the mechanics of achieving pollination need to be precise. But what are they? Ants, beetles, flies, fruit flies and white-eyes (sun-birds) have all been suggested as agents, but whether or not it is the smell of the quickly-rotting flower that attracts them is unknown. For successful fertilization of the seed ovule, the smelly stage may already be too late.

The methods of seed dispersal are almost as debatable. In order that the filaments of the germinating seed may infiltrate a *Tetrastigma* vine successfully, the vine must be damaged in some way; this was established by experiments in the Botanic Garden in Bogor. On the forest floor, this is most likely to be caused by the sharp hoof of a large deer or the horny nail of an elephant's foot. Arnold recorded that the area in which he found his flower was 'covered with the excrement of elephants'. If an animal has previously walked over a *Rafflesia* flower and is carrying on its foot some fertile seeds, then it is possible some might be implanted in a new vine, which the animal damages slightly by stepping on it.[1] This sounds far-fetched, perhaps, but no one has yet thought of a better explanation, and the flowers do seem to occur in big game areas and, conversely, not to occur in places from which they were formerly reported when these are no longer frequented by the larger hoofed mammals such as elephant, deer and tapir. It is a reasonable inference, therefore, that the protection of the one somehow involves the protection of the other, even though the exact method of dispersal has not yet been fully worked out. In Dutch times, several '*Rafflesia* Monuments'

[1] Overhead-growing *Rafflesia* plants can be accounted for by the growth of the filaments through the trunk of the vine after implantation. Dr W. Meijer has suggested 'aerial implantation' by pecking birds or sharp-clawed squirrels, but if this were true, some flowers should break out much higher up on the vine than the current known maximum of 4 m from the ground.

were established in western Sumatra which demarcated as a reserve only a few hectares in the vicinity of known flowers and host vines. Many of these are now near human settlements or in hunting reserves; the animals have been shot or have moved away, and a recent investigator reported that he could find no trace of the *Rafflesia* plants either.

12

The Climbers or Lianas

THE presence of so many woody climbing forms of plants is one of the most noticeable features of tropical forests. Their long water-bearing stems cling to the tree trunks or hang in loops between one high branch and another or, in their old age, lie in great weighty coils upon the ground. No writer or artist describing a 'jungle' has ever left out the snaky stems of the climbers or lianas, nor has any jungle dweller remained unaware of their flexible usefulness. Even in films, the camera lingers on these ropes of the forest. Breathless, we admire Tarzan as he swings halooing through the trees or sails across some incredible chasm hanging on the end of a thick vine; or, variously, we are filled with horror when our hero grasps a climber we know is really a deadly serpent, so sleek are the skins of both. Climbing plants also add to the jungle's mystery. They seem to confuse the pattern of the rain forest, supposedly all tall towering trees, by criss-crossing between them and making us, at first glance, mistake their limbs or leaves for those of the tree itself. Or sometimes they bar our way, tripping or entangling us. In other words, when we think of a 'jungle', it is a jungle full of climbers we are thinking of, a dark impenetrable maze of vegetation.

For the most part, this is a mistaken idea. Undisturbed rain forest has wonderfully clear ground between the trees. The climbers themselves are thrusting eagerly upwards towards the light, doing little if any harm to adult trees because, unlike parasitic plants, they are merely using their hosts as supports. Their fruiting and flowering can only take place in the canopy; to regenerate, they must reach the sunlight too.

Altogether, climbers comprise about 8 per cent of the total number of plants in the rain forests on land between sea-level and 500 m, which is also the type of rain forest richest in hardwoods and therefore the most valuable commercially. For the climbers, this is a pity. Invariably, the lowland forests are the first to be exploited, and the lianas, which are regarded by commercial foresters as a nuisance, are hacked and chopped away under the notion of 'habitat improvement', or are inevitably felled at the same time as the trees on which they have depended for support. From the forester's point of view, their presence certainly adds to the damage caused by logging since they tie the crowns of several trees together, either holding up the desired tree, or bringing down more trees than intended. That these tangled and sometimes incredibly strong coils of vegetable matter might be of value in themselves has so far played little part in tropical forestry management.

Nor have they received much scientific attention. As an important component of the flora of an area, they are usually ignored. In 'species counts', they escape the net of such categories as 'all trees with a girth of over 10 cm' or 'all angiosperms' or 'all vascular plants'. The number of species per hectare can often be doubled if climbers, creepers, herbs and epiphytes such as orchids are counted as well as trees. Climbers are also difficult to study. If an area has been clear-felled, the homeless vines lie in a mess on the ground, almost impossible to sort out. On the other hand, it is often necessary to cut down one or two trees to obtain a climber's fruit or flowers, and even with much care it is likely that these tender fast-growing parts will be damaged. If samples of the stem wood are to be taken at various levels, which is important, this requires much patient tracing of the climber in question if, as is usually the case, there is more than one climber in the tree. It is not as if the leaf shape were of help to one here: climber leaves usually vary in form and size depending on their distance from the ground.

Nor has much work been done yet on the ecology of climbers, for instance, on the contrast between a forest with many and one with few or none. What role do climbing plants play in the rain forest? Certainly they help to form above-ground pathways for arboreal animals, strong enough even for the heavy apes to use. In this role, they aid animals in their search for food and aid plants by the resulting dispersal of their seeds. Also, they create more litter, and therefore humus, on the forest floor proportionally than the trees which support them. In an African rain forest, it was found that while trees produced 60 per cent of the litter, the climbers, which had only 5 per cent of the weight in wood of the trees, accounted for very nearly all the rest (36 per cent)! Quickly recycled in the rain forest by fungi, bacteria and insects, litter provides most of the nourishment available to the plants; little ever remains in the soil. But a steady degree of moisture and heat is essential to its rapid

I Ficus variegata (cauliflorous fig)

50 cm

II Calamus ornatus (rattan)

III Eugenia aquaea (jambu)

IV Myristica fragrans (pala), nutmeg

100 cm

Va Nephelium lappaceum (rambutan), a tree in fruit

Vb Nephelium lappaceum (rambutan), detail of a branch with fruit

4.5 cm

18 cm

VI Durio zibethinus, durian

VII Artocarpus incisus (kluwih), a breadfruit tree

VIII Climbers in the rain forest

IX Alyxia alata (pulasari)

X Cymbidium hartinahianum, a newly discovered ground-flowering orchid

5 cm

XI Pinus merkusii, leaves and cones (male with pollen)

decomposition, and by linking the tops of trees climbers help to control these conditions, which any opening in the canopy disrupts. When a gap occurs in the canopy, for example, when an old tree dies a natural death and falls, climbers seem to hurry in from all sides to close it with a green curtain. This is a growth opportunity for them, and at the same time they provide the shade necessary for the old tree's progeny to grow.

Other species of climber—hardy, sun-tolerant ones—help equally well in this respect when they invade secondary vegetation. In this case, they contribute to the gradual and natural replacement of 'pioneer trees', whose seeds are only able to germinate in the hot sun of an exposed landslip or abandoned *ladang*, by the richer vegetation of the rain forest. Rain forest plants are without exception extremely sensitive to minute variations in humidity and heat; under a closed canopy, as anyone who has walked in the rain forest knows, the temperature at ground level is often 10 to 15°C less than on ground exposed to the full force of the tropical sun.

Though climbing forms of plants occur in most plant families, nearly all of them are confined to the tropics. For this reason, because tropical botany is still relatively underinvestigated, only a small fraction of these plants are actually known to be useful to man, and this fraction only by people living in or near the forests who have this knowledge handed down to them by local traditions. The single and very important exception to this are the rattan palms, on which the material culture of the South-East Asian nations is partially based (see Chapter 3). Local uses of other climbers vary from area to area but commonly include fibres for string and rope making; bark cloth and paper; sweeteners for food (one species contains a substance 1,500 times as sweet as sugar—see Paijmans), insect repellants and fish poisons (derris root powder), and valuable additives to the diet, the seeds of some climbers being rich in Vitamin C and the leaves of others containing calcium. Where their nutritive value has been recognized, they are collected regularly, for instance, in New Guinea, and in Borneo by the Dayaks.

Of most potential economic interest, however, are the medicinal values of many climbers. Though this is still a relatively unexplored field, certain families are known to have a high proportion of medically important species. The *Apocyanaceae*, for instance, include *Strophanthus*, *Rauwolfia* and *Vinca rosea*; *Vinca*, the common periwinkle, is used in the treatment of certain cancers. The genus *Chondro-*

dendron of the *Menispermaceae*, which are mostly climbers, is an important source of curare, the famous South American Indian arrow poison which modern doctors now use in minute quantities to relax muscles and reduce nervous tension. The *Asclepiadaceae*, *Conneraceae* and *Dioscoreaceae* are all families which contain medicinal plants too, and extrapolating from the number of known species, one can estimate that probably about 4 per cent of Malesian rain forest plants have medicinal value. With several hundreds of plant species in a hectare of lowland forest, each hectare could contain as many as twenty medicinal plants about which we as yet know little or nothing, and which may also soon disappear, particularly if they are rare, before they have been described scientifically and certainly before their uses are appreciated by scientists and doctors.

One local climbing plant, however, is already well-established in the pharmacopoeia of Indonesia: *pulasari*, a name given to several species of the genus *Alyxia* (family: *Apocyanaceae*), which are in common use, perhaps 6 out of the 57 growing in the Malesian area. Like the other members of this family, *Alyxias* exude much white latex when cut; the best known, but not medicinally used, relative is *Dyera costulata*, the *jelutong* tree, whose latex is the basis of chewing gum.

Alyxia climbers range in length from a few metres to 25 m or more, and the leaf shapes and arrangement also varies, but the genus as a whole is easily recognized by two obvious characteristics: its linked fruits, up to four in number, and its distinctive perfume, coumarin, a crystalline substance also found in such plants as the Tonka bean of Guiana, in South America; to Europeans, the smell is that of fresh-mown hay, green and sweet. In Indonesia, the aromatic property of *pulasari* is used to disguise the bitter taste or less attractive smell of the medicinal herbs with which it is combined, and it is one of the most important ingredients of various types of *jamu* or herbal tonic mostly for this reason; the quantities of tannin and alkaloids it contains are too small to make it of much medicinal value, unlike its more important relatives.

The various kinds of *jamu* in Indonesia are regarded, particularly by the ladies of Java, as rejuvenators, vital after childbirth and healthy as tonics throughout life. Even after many children, good *jamus* are said to keep a woman youthful and vigorous, or at least beautifully languorous, and above all attractive to her husband. But they must also be taken daily to have any effect, and the business opportunities this offers have not been lost on several astute ladies of central Java and their imitators, nor on the veritable army of youngish women who hawk the tonics through the towns and kampongs. But the ingredients of the best *jamu* recipes are carefully guarded secrets known only to the direct descendants of the original proprietor, except for *pulasari*, revealed by its distinctive perfume. In *Alyxia*, the perfume seems to be concentrated in the bark, which is removed from the stem of the climber and sold in packets in much the same way as cinnamon bark.

13
Partnerships of the
Rain Forest

MUTUAL dependence between plants and animals or other organisms is common in the rain forest. The trend among temperate climate plants has been away from any obvious relationships that limit dispersal and adaptability, but tropical forest species are still enmeshed in webs of interdependence vital to their mutual survival. A great many of them rely on animals for either pollination or dispersal or both, the mutual adaptation of fig to fig wasp being an interesting example (see Chapter 2). There are also plants which live on, with or within other plants, in symbiosis. Symbiosis is often an adaptation to nutritionally poor conditions and is, usually, of equal advantage to each partner. Still other plants depend on the existence of another, much 'lower' organism for at least an element of their survival, as the leguminous trees do on bacteria or the dipterocarps and orchids and numerous other rain forest plants do on their root fungi. The partnerships of the rain forest are both important and complex.

In Chapter 2, I tried to describe the exact relationship between a species of fig and the species of wasp which pollinates it. Such precise specializations in the shape of the flowers, matched by others in the form of the wasp, plus such matched sequence of maturation imply simultaneous and complementary evolution, with the result that little fig pollen, valuable genetic material, is wasted. This has the virtue of economy, good for plants as well as for people. But the disadvantage of such specialization is very quickly apparent when a plant and its pollinator are separated, by forest destruction, for example, or by attempts to grow the plant exotically. Sumatran figs in the Bogor Botanic Garden have never borne fruit because their particular wasps did not accompany them.

Fig trees are the food producers of the forests. Plentiful, prolific, they are heavily relied on for survival by birds and other animals; correspondingly the figs can rely on the animals to disperse their seeds. In fact, the fig fruit could be described as the ultimate in 'dispersal mechanisms'. It is sweet, satisfying and easy to eat. Although packed with seeds, they are too small to interfere with the fruit's consistency—rather, they add to it—while at the same time they are hard enough to pass uncrushed through any digestive tract. This they soon do because of the fig's well known laxative effect, which in turn ensures the new seedling a good start in life.

Of course all fruits are by intention 'inviting'. From an evolutionary point of view, the enclosing of seeds in edible pulp could be described as a device on the part of immobile plants to enlist the aid of mobile animals in seed dispersal. Wide distribution ensures survival, for plants as for other organisms. On the other hand, it is also important for the plant's survival that its fruits are not eaten before the nut or seed is ready for germination. To this end, unripe fruits are usually camouflaged by colour, green, and protected by chemically-created bitter or sour tastes, or by an abundance of latex, as in figs. Eye-catching colour, usually red, attractive, or at least strong, scent, good taste and sufficiently soft consistency without unpleasant juices or sticky gums, are attained only when the seeds inside the fruit are fully developed and ready for planting.

But unless the fruiting species produces abundantly and unless the seeds distributed can count on the presence of good growing conditions over a wide area, it may be that animal assistance can become a threat rather than an aid to survival. Providing an abundant source of food will lead to an increase in the number of animals and overcropping may result. On poor soils especially, it might happen that the immobile tree cannot produce sufficient fruit to feed a large population of animals or that its possibilities for dispersal have already reached the maximum for the prevailing conditions. In this case, it might flower and fruit erratically, avoiding animal associations and hoarding its reproductive strength in order to react to the stimuli of such climatic conditions as only occur every few years. It is tempting to speculate that some such evolutionary process, involving survival, might be the reason for the erratic flowering and fruiting of the dipterocarps.

Plant–animal relationships are not confined to the pollination of flowers and the dispersal of seeds. There are many interesting associations between ants and plants in the rain forest, occurring most frequently amongst epiphytes, which use another plant for

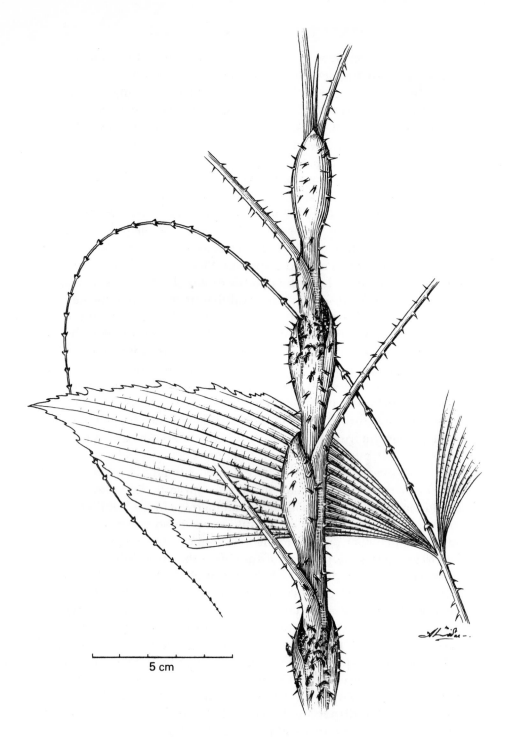

5 cm

Fig. 20   *Korthalsia scaphigera (rotan merah)*

support and have to rely for their water and food on aerial resources, or among plants growing in poor soil conditions, which may explain the presence of so many ant plants in hilltop forests. There is quite a long list of plants with resident ant populations, which includes several species of the *Euphorbia* and *Rubiacea* genera and three species of ferns, as well as rattan, *Korthalsia scaphigera* (Fig. 20). The ants are housed and sheltered in the plants' hollow stems or twigs, which appear swollen because of the network of cavities in them, or, in the case of epiphytes, in the swollen tuberous roots. Fig. 21, of the epiphytic *Hydnophytum formicarum*, a *Rubiacea* is drawn to expose the galleries or cavities made by the ants inside its tuber. Within such galleries, the ants often 'farm' large 'herds' of scale insects, which suck the sap of the inner surface of the twigs or tubers, while the ants feed off the secretions of the insects. In these cases, therefore, the plants indirectly feed as well as house the ants. But how do the plants benefit by this arrangement? This is not so clear. Some ants do carry in and/or discard such debris as pieces of leaf, bark or rotten wood and the indigestible parts of insects, which is of no use to them as food but which, once decomposed, would nourish the plant. Ants would also, presumably, eat any caterpillars feeding destructively on the leaves, and are known to attack fiercely any animal breaking off twigs or branches. The *Korthalsia* ants are said to be particularly aggressive, which makes collecting this rattan even more tiresome than rattan collecting usually is.

Ant plants which are associations of mutual benefit must, by the way, be distinguished from plants which actually *feed* on ants and other insects: the *Nepenthes* or pitcher plants, for instance. Pitcher plants, common on poor sites such as damp cliff sides or moist areas in old volcanic caldera, catch insects in the greatly enlarged ends of their climbing tendrils. The plants produce a digestive liquid retained in the base of the pitcher; insects, attracted by the sweet secretions of glands situated near the opening, fall into the water, cannot climb up the slippery inner surface of the pitcher, and soon drown; the plant is then able to absorb the nitrates released by their decomposing bodies.

Decomposition of vegetable or animal matter into food for plants is mainly the province of bacteria, assisted in some cases by fungi. Their role is all-important in the rain forest where re-use of the proteins and minerals temporarily locked in the bodies of insects or the vegetable tissues of plants is an urgent matter. Bacteria are usually associated with disease, but these micro-organisms are

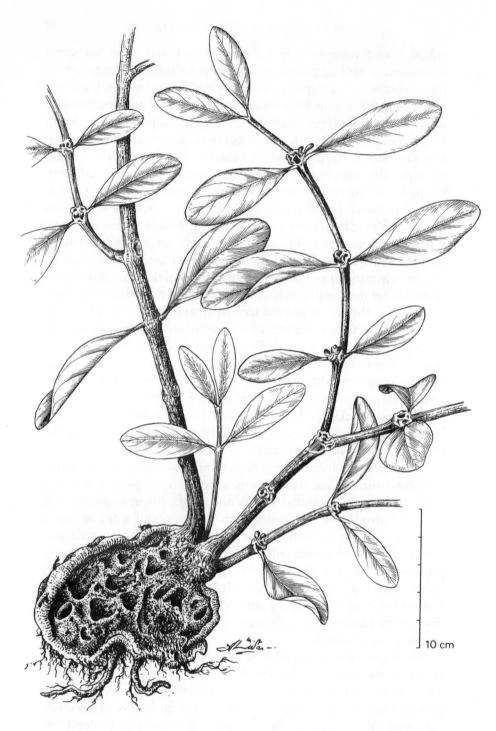

10 cm

Fig. 21 *Hydnophytum formicarum*

actually at work everywhere decomposing and reprocessing waste matter either into soluble salts, which nourish plants, or into nitrogen gas, which replenishes the vital supplies of nitrogen in the earth's atmosphere. They also break down food for absorption by the body. They are nature's recyclers, and all living matter depends directly or indirectly on their existence.

Some plants, however, have special relationship with certain bacteria, notably members of the *Leguminosae* or pea family. Leguminous plants harbour bacteria in nodules on their roots which are actually able to manufacture nitrogen compounds (salts or nitrates) from the air. The consequent liberating of extra nitrates into the soil around their roots is of great benefit to the plant, particularly in poor soils. This is why, in agriculture, so many legume species are used as 'green manure'; they both cover and improve the soil. This attribute is equally helpful in the rain forest. Many rain forest soils are 'lateritic' or 'feralites', that is, red-looking earth from which the alkaline compound, silica, has been leached by the constant high humidity and rainfall; this leaves an acid and clay-like soil with a high content of aluminium and iron compounds. Without the beneficial effect of normal bacterial and fungal activities usual in the rain forest and the further assistance of nitrogen-fixing nodules, it is doubtful if such giants as the *tualang* (*Koompassia excelsa*) and the durable heavy timber of *merbau* (*Intsia palembanica*) would grow on such organically poor soils.

Fungi comprise the large order of 'lower' plants which includes mushrooms, moulds, rusts and yeasts. Fungi lack chlorophyll, the catalyst or agent which enables most green plants to manufacture their own food from inorganic materials by using the heat and light of the sun. Fungi must therefore obtain their nourishment from dead or decaying matter, become parasitic on a host plant or live in symbiosis with another plant form. In the first case, when fungi live off dead or decaying matter, they contribute to its breakdown into plant food. In the case of parasitic fungi, where the host plant derives no benefit, it does not follow that the fungus will actually harm its host, though many do. But there are some relationships with another plant in which the fungus plays such an important role that the relationship must be called truly symbiotic.

Lichens are an outstanding example of symbiosis. They are entirely composed of fungi and algae in about equal proportions. The fungus controls the shape of the plant, provides the mechanism by which it is attached to the rock, soil or bark, and supplies water for both organisms by absorbing it from the air through the hyphae, or thread-like structures, of which it consists. On its part, the alga involved manufactures the organic food for both, although some mineral salts are also released from the surface to which the lichen is clinging; this is caused by the acidic action of the fungal hyphae at the point of contact. Lichens are therefore important colonizers of rocky bare surfaces since they contribute to their breakdown into new soil.

Another symbiotic relationship in which fungi play a part is in the association between a fungus and the roots of a higher plant known as 'mycorrhiza'. Mycorrhizae often occur in heathland plants and in pines and beeches, all of which typically act as pioneers on exposed sites. In such plants, the mycorrhizae seem to replace the hairs on a root which normally function as water-absorbers, and it is likely that under harsh conditions they are more efficient at the job. Some pines are quite unable to grow without their fungi, and this fact has to be allowed for in reafforestation and plantation work. In rain forest conditions, however, mycorrhizal associations are equally important, and many plants have them, the dipterocarp family being a good example, and fungal hyphae actually penetrate the roots of orchids, so that the life of one is inseparable from the life of the other. In enabling these plants, among the largest and the smallest of the rain forest, to make the best use of a fragile and transitory system of nourishment, mycorrhizae have probably contributed much to the success of these two families.

14
The Orchids

MANY people consider orchids to be the most beautiful of all flowering plants, offering to the senses an endless variety of, colour and intricate form, and to the collector the excitement and challenge of acquiring rare species from inaccessible jungles, or of breeding ever more elaborate hybrids. More lives have been lost collecting orchids from their wild habitats, and more money spent, than in the quest of any other genus of flowering plants; in the process, so also have many orchid species passed into extinction in the wild, and many more have become seriously endangered. The notion that human beings always mistreat what they love best seems true enough in the case of many orchids, and commercial greed has certainly played its part in what has often been shocking and irresponsible exploitation.

Unfortunately, some abuse still continues. Hired collectors, operating from boats on the big rivers of Kalimantan which give access to the far interior of the country, have been seen tearing down all the orchids they find; they then take them to the coast, for their employer to sort out the wanted species, throwing away the rest. By this method—employment of cheap unskilled labour and indiscriminate collection—it is very much easier to collect orchids from the forest than to grow them from seeds. Most of the exporters in Indonesia certainly do not attempt orchid cultivation, which can be both complicated and time-consuming, but they could, and should, take better care of the plants they do receive. Since most of the low-altitude orchids are either commonly available amongst orchid fanciers in cultivated form, or are completely extinct, collecting is now done not only far from the coast but at high altitudes as well, without, however, any effort on the part of the exporter to keep high-altitude plants in montane conditions until sold or exported. Many orchids therefore die before sale and the vendor obviously tries to replace them—again from the wild.

Airline companies also contribute to over-exploitation by offering cut-rates for large export consignments, which encourages over-collection and increases the risk of mortality. It also makes it much easier to smuggle out officially protected endangered species. As in the bird trade, the customs inspector has to be an expert to detect the presence of a few rare species mixed in with a lot of common ones, especially if the rare ones are not in flower at the time. Protected orchids do often leave the country of origin in this manner: witness the many advertisements in supposedly reputable orchid-fancier journals which offer 'rare jungle plants'. One begins to wonder, probably with good reason, whether or not a dealer in orchids much cares about the extinction of a species in the wild, since this would only serve to increase the value of the few he has or of his nursery-raised varieties.

Orchids also suffer, as do all wild plants with special growth requirements, from the steadily increasing destruction of wild habitats. The orchids of Singapore are a good example. When the plants of Singapore were first collected and named by scientists in the early 1800s, the flora native to the island numbered about 2,000 species, of which 200 were orchids, and these, as is usual in the rain forest, were mostly epiphytic in habit. Epiphytes, or plants which use other plants for support, are particularly vulnerable to any change in their habitat. If their support is cut down, of course they die too, but they are also very sensitive to even quite small variations in the humidity and temperature of their environment since they are totally dependent on their own water storage systems, such as the enlarged tubers of some species which provide a sufficient supply for the habitat to which they are adapted. If, suddenly, far more sunlight and wind are let in, epiphytes wither almost instantly, and indeed are never found growing in both exposed and dry conditions. In this connection, it is sad to see orchids and other epiphytes tied to posts in the hot sun of some village street, even if their roots are embedded in porous tree-fern stems. It is an unnatural environment for them, and when they die it only means that more wastage has taken place in the name of some mistaken scheme for village beautification. The orchids of Singapore, however, were not even transplanted but only casually destroyed: according to Holttum they were most plentiful in the mangrove and freshwater swamp along the coasts; with Singapore's rapid development, they were the first plants to disappear.

Not only are the epiphytic orchids highly sensitive to changes, their growth in the wild or in cultivation absolutely depends on the presence of certain species of fungi whose hyphae or filaments penetrate the root of the orchid and, apparently drawing on decomposing matter in the host tree's bark, supply the orchid with vital food for growth. These fungi are quite common in bark and other vegetable tissues of rain forest plants, fortunately, otherwise the tiny, light, almost foodless seed of an orchid would never succeed in developing. This relationship is actually maintained throughout the life of the orchid, but is especially important at the start.

Orchids are also more vulnerable to change than many other plants because they are so extremely slow-growing. The time a seedling takes to reach maturity can be shortened artificially, that is, under laboratory conditions, but the life cycle of most wild species takes four years to complete. Though this is nothing like the lengthy regeneration cycle of the dipterocarps, where half a century can pass between a seedling's germination and the tree's first flowering, it is still very slow for a non-woody plant. This is an important factor to remember when trying to calculate possible regeneration rates under natural conditions.

Against these rather negative characteristics, or orchid-debits as it were, one has to balance several interesting facts. First, the orchid family *as a whole* is huge, consisting of some 18,000 species or more. Secondly, their distribution is almost world-wide: their ecological range includes all but the most extreme climates and conditions, such as the sea, the driest deserts and Antarctica. Thirdly, genetically-speaking, orchids are rather promiscuous, that is, although many orchids have their specific pollinators, cross-fertilization between species often produces fertile hybrids even under natural conditions. Since in a laboratory cross-pollination is easy, this last fact makes orchid cultivation less difficult than it might be and also accounts for the almost endless proliferation of hybrids which is such an important part of the huge orchid-growing and breeding industry.

Furthermore, the popularity of orchids among professional horticulturalists and the wide spread of orchid-fancying among amateurs has without doubt led to much of the excellent legislation which has been passed to protect rare species. Of course, greed for the new and the rare has not lessened and, as usual, new legislation cannot keep pace with new botanical discoveries, but the fact remains that orchids, as a family and in theory, are the best protected of all plants. The continued existence of the rarest species in the wild—about 100 out of the 5,000 or so species recorded for the Malesian area—is probably now more a question of law enforcement and of controlling the progressive destruction of their habitats than of preventing the 'ignorant rape' of the forests which characterized the last century.

15

The Rain Forest by Contrast:
Pinus Merkusii, a Tropical Pine

OF course Malesia is not all rain forest, low-lying, ever-wet and covered with huge trees; other habitats of many sorts exist. Besides the drier seasonal 'monsoon' type of forest, there are also large areas of savannah grassland—the 'destruction stage' of the monsoon forest—in eastern Indonesia, and extensive mangrove swamps along the many sheltered coastlines from Sumatra to New Guinea. There are also interesting 'sand seas', 'ash-lakes', and 'mud-streams' (*lahar*), all the direct result of volcanic activity, which are colonized by special plants; and, in the highlands of Sumatra, there are even indigenous pine forests, the *Pinus merkusii* being the southernmost extension of this characteristically northern temperate climate genus.

Visitors from northern temperate zones take the presence of pine forests for granted, so common are they in Europe, northern America and northern Asia; but in the southern hemisphere and particularly in the tropics, they are unusual, unless they are introduced species. In fact, the division of the plant world to which pines and all other conifers belong, the gymnosperms, once the dominant plant type found all over the world, are now represented by relatively few families and genera, and although some of these are so successful they cover vast stretches of land in the northern hemisphere, they are rare here in the tropics where they (and most other plants) originated.

Because they were forced to move out of the tropical centres of evolution by the more advanced, more complex and more successful angiosperms, conifers, particularly the pines, gradually acquired characteristics through the process of natural selection which enabled them to survive in ever less favourable climates, and to develop outstanding abilities to colonize areas where other trees could not grow. They developed the thin, hard leaves commonly called 'needles' which can withstand both drought and cold, and tiny, light seeds which are carried easily by the wind and can germinate in dry, bare soils; strong sunlight aids rather than deters this process. In fact, pine seeds are unable to germinate under cover, even under the parent tree's cover, and thus must find new bare earth in which to grow. They have evolved into true pioneers.

Where they do grow, they grow strongly and quickly, aided by the condition of 'fungus-root' or mycorrhiza which all pines have and which replace the usual root 'hairs' as water- and nutrient-absorbers. Being quick-growing, pines are an ideal crop to grow for fuel, or for matchwood, or for making paper pulp, and because of their single-stemmed shape, they provide easily lumbered timber for house-building, furniture, crates and so on. It is easy to see why pines are so popular in plantations, and the qualities that make them so are the opposite in every way of those of a rain forest tree.

But the tropical pines do have an important role to play in the rain forest. The pines' ability to grow on bare surfaces is one of the ways the rain forest is able to heal its 'wounds', whether these are natural, like landslips or mudstream avalanches, or man-made, like the results of a swidden or slash-and-burn type of agriculture. Wind-carried pine seeds germinate well in such conditions. The pines grow, the earth is covered and shade is provided. Then, when the pines have colonized the area completely and their seeds will no longer germinate in their shade, shade-loving species will grow, gradually taking over when the pines die; more gradually still, the true rain forest species will move in, the trees whose large, slowly dispersed seeds cannot tolerate heat and light, and which need constant moisture and cannot wait, dormant, for the right conditions in which to germinate. This process by which the rain forest returns is called by the botanists a 'natural succession'. Typically it is the pines which will start such a succession in mountainous areas. At low altitudes, or in places where there is no dryish period of the year, pines are not able to act as colonizers; this role is left to other light-loving species which are collectively known as *belukar*, or bush, or secondary vegetation.

Either method of regeneration is a means to an end as far as the rain forest is concerned, and the difference between them would be unimportant were it not for what is so often the capricious hand of man. *Belukar* has little or no commercial value. Were it not for increasing pressure on the land by people who for many reasons are forced to make more frequent use of ever decreasing areas of natural forest, the natural succession which starts with *belukar* would probably proceed normally. But with pines, the situation is different. The 'hand of man' here is not called 'shifting cultivation' but rather 'cost-effective timber production'. If the slopes on which the pines grow are a little too steep, or the altitude too high for convenience, the difficulties are far outweighed by the value of the cash crop to be obtained. All those fine straight trees growing so close together! To a person who knows how to convert wood into money, pines are the ideal tree. Of course, cutting down the pines spoils the natural succession just as surely as the too hasty fire of the cultivator, but it is much more tempting: more pines will grow, and can be cut again. From cutting down natural stands, it is only a simple and apparently logical step to extend the area of pines by artificial means: by plantations. Pines can be planted instead of the 'useless' natural oak and beech-dominated forest of the upper mountain slopes, and little or no thought is given to the erosion caused by frequent logging of steep slopes, or to the eventual 'podzolization' of the soil which is the result of continuous pine cover: the pines' highly acid leaf fall inevitably makes soil itself more acid (also discussed in Chapter 10 on the *Agathis*).

Pines are also increasingly being planted at lower altitudes, even in lowland Kalimantan, where they are used for the reafforestation of clear-felled timber concession land. But this has not been at all successful. In such areas, they do not seed properly because their pollen-forming ability depends largely on climatic stimuli, and in these conditions the trees also have an unfortunate tendency to 'fox-tail' or develop a long leading shoot devoid of branches (Figs. 22a & 22b); this interferes with its normal development into a strong, straight tree. Also, pines grown at low altitudes cannot survive any competition from species more at home there, which means that for the first few years at least such plantations need continuous and expensive weeding. But once again, the tempting profits pine plantation owners dream of making seems to outweigh all these disadvantages.

In all this increase of land under pines instead of under natural forest, the *tusam*, *Pinus merkusii*, has been the main species involved. There are, in fact, only two tropical pines in this part of the world: *P. merkusii*, named after a governor of the Netherlands Indies, the native of Sumatra, and the *P. kesiya*, native to the Philippines and certain parts of the continental South-East Asia. The other well-known tropical pine is *P. caribaea*, from Central America. Of the two

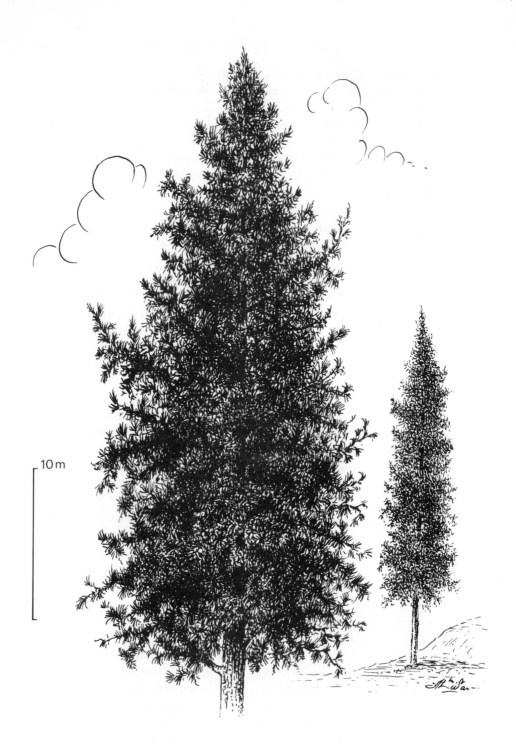

Fig. 22a   *Pinus merkusii* (*tusam*), a healthy tree

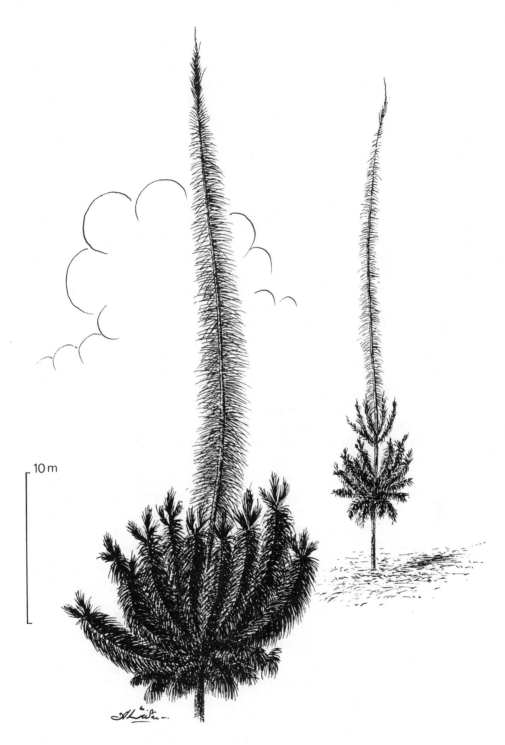

Fig. 22b *Pinus merkusii*, a sick tree showing 'fox-tailing'

tropical Asian pines, the *tusam* is the most popular species for plantation work in Indonesia primarily because its seeds are readily available; also, as a native, its use by both local and foreign timber concessionaires in replanting experiments neatly avoids their being criticized for planting 'exotics'. Changing the proportion of one native species to another is much easier to defend than the introduction and widespread planting of a foreign species, whether the subject is trees or theories or even people. Besides, quite extensive areas of these pines were common in Sumatra at the beginning of this century, and were long thought to be the natural vegetation of parts of the province of Aceh and of the Tapanuli area. It was not until the Dutch tried to improve the quality of these 'natural' stands, in order to collect the trees' sap for making rosin and turpentine, or for timber, that the role of fire in spreading these trees was appreciated. Like the hardy montane casuarinas of East Java and a few savannah species, once these pines are about ten years old, they are resistant to fires which would kill other trees. Herdsmen wanting to promote grass growth and still retain useful timber can therefore 'fire' large areas, spreading the grass and keeping the natural forest at bay. When fire is used frequently in this way, pine-dominated grasslands, in which only rarely is any other species of tree to be found, soon appear to be the natural vegetation of an area—until the role that fire has played in their formation is understood.

But neither the *tusam* nor the *cemara*, the montane casuarina, *Casuarina junghuhniana*, can resist too-frequent firings, and the fires which clear the land and promote the growth of these pioneer species can equally well destroy them. In some ways these trees themselves contribute to their own destruction by the amount of dry combustible litter they shed upon the ground, so that accidental as well as deliberate fires frequently rage through such forests. It is amazing how much fire such trees can stand, particularly the *cemara*, which will even sprout from a charred stem or a damaged root; but eventually fires can destroy all the sprouts and seedlings and impoverish the land to the point where only tough grasses with ground-creeping root systems like those of *alang-alang*, *Imperata cylindrica*, will grow. Once *alang-alang* takes over the land, no grass or shrub, no matter how much of a pioneer, can find a root-hold. This sort of succession, far from natural, is all too familiar in South-East Asia, where large tracts of land are now covered with this sharp-bladed grass, inedible (unless very young) and seemingly ineradicable. Such wastelands are the end result of man's interference with nature in the tropics which has started with fire and is then continued with its use, either deliberate or accidental.

The extensive planting of pines must therefore be regarded as a risky use of tropical land, tending towards its eventual degradation, because inevitably the pines will be regarded as the end crop and the next natural succession stage will be prevented, either by fire, crudely, or by a sophisticated and profitably short cutting cycle. Because the latter works in temperate zone plantations in different soil and under completely different climatic conditions, pine planting has become a cliché of European or American forestry practice. In Malesia, pines may have a role to play in converting wastelands into useful temporary plantations, but they are not a long-term solution. Such forests, apart from the fire risk, provide no food for either men or animals nor can they revive the soil's fertility, and therefore can never be a substitute for the broad-leafed species of the rain forest.

Epilogue

A NOTE ON TROPICAL BOTANY

As a science, the study of plants is not very old, perhaps 450 years old at the most. Botanists are still busy classifying and rearranging; many of the families of the flowering plants, for instance, of which there are an estimated 250,000 species, are still in flux, and every year brings new 'revisions' from the busy taxonomists. The reason for this is that two-thirds of the flowering plants in the world are found *only* in the tropical regions, and it is really in the last fifty years that tropical botany has been properly studied. That is, until relatively recently, the majority of scientists and scientific organizations, with certain famous exceptions, were based in temperate climate countries. Formal, written botanical knowledge, the body of 'established fact', the field studies and the other devices whereby professors of botany passed on knowledge to their students, were all based on the plants of the temperate zone, poor in number and restricted in variety.

Imagine a student who has only been taught temperate climate botany suddenly put down in the middle of Borneo. He would be amazed at the strange methods of flowering, confused by the look-alikes of different species, and bewildered by the immense variety. In the whole of Great Britain, for example, the total number of plant species is about 1,480. But in Java, with half the land area and nearly twice the number of people, there are still reckoned to be 4,500 species! The Malay Peninsula has 8,000; New Guinea about 9,000; the estimate for Borneo, which is thought to have the oldest forests in this part of the world, is over 10,000. To a botanist, that is the definition of a rich flora. Compared to Borneo, Britain is a desert.

Tapping this vast reservoir of botanical knowledge systematically is still far from complete. Basic data on many species are still missing or very limited, and much comparative work remains to be done. Since, as Professor Richards (1952)[1] said, 'the tropical rain

[1] P. W. Richards, *The Tropical Rainforest*, Cambridge, 1952.

forest is the home *par excellence* of the broad-leaved evergreen tree, the plant form from which all or most other forms of flowering plants seem to have derived', modern botanists have come to realize the enormous importance of having comparative data from the tropics in any study they may undertake. They now see, for instance, that unless evolutionary theories are tested in the tropics they are hardly tested at all; that plant form cannot really be studied unless the full range of such form, which is only found in the tropics, is understood; that adaptations studied without reference to tropical responses are being studied in a kind of temperate vacuum. In short, that the biology of plants will never be fully understood until the botany of the tropics has been extensively and intensively explored—in full.

The only trouble is—time. Botanists see the material they need to, and must, study disappearing faster than they can work. For some species, and no one knows how many, it is already too late.

THE ISOLATED FOREST DWELLERS: HUMAN LIFE IN HARMONY WITH THE LIFE OF THE FOREST

Until now, and especially in Indonesia, there has been enough remote and undeveloped land for small groups of people to follow their traditional ways of life without interference; and since such people rarely make any drastic change in their environment, their life is often life in the rain forests. The forest provides their food (wild vegetables, fruits and hunted animals) and their material culture (houses or shelters, boats, hunting equipment, twine, rope, poisons and medicines). There are reckoned to be over 4,000 plant species used by forest dwellers as food and medicine alone, many of which are local or endemic, known only to small groups whose knowledge of the forest is passed on orally, from generation to generation. Adapted to life in the forest, self-sufficient in it, using its products but never destroying their source, hunting forest animals but only according to need, such people both protect the forest and are protected by it. Isolated until now by vast tracts of 'jungle', they have managed to live in balance with nature, in harmony with the lives of the plants and animals with which they share this special environment.

This balance is not achieved casually, without effort. Exploitation in any commercial sense is unknown or only at the very low level of barter, and such peoples operate at subsistence level. Life is hard, requiring almost constant work from every member of the community; the arts of leisure have no place here. Such people also, to survive, have had to overcome the main problem besetting modern, urban, 'civilized' man: his terrible fertility. This is not an automatic result of hard living and lack of medical skills and drugs; on the contrary, many fertility controls are built into such cultures: sanctioning the use of abortion, condoning infanticide, abstaining from sex during weaning, hunting or other extended ceremonial periods.

Environmental controls or sanctions are also very much a part of such cultures. If trees have spirits, they are not cut idly, for mere profit, but only out of real need for the timber: a house, a new boat. Nor are animals killed for the pleasure of the hunt; their spirits too must be placated, their flesh needed. Such restraints are cultural expressions of the idea, the belief, that plants and trees have as much right to life, and the space in which to live, as people do. Little of this attitude seems to remain in 'modern' life.

Even the much reviled system of shifting agriculture or *ladang* making (which we may discover, finally, to be the only agricultural method applicable to rain forest soils) works well at the 'primitive' level, when the necessary rules are followed. There are examples of this throughout South-East Asia and the islands: when the local *adat* is strong, the vital fallow periods are observed, the field size is limited and the forest returns to claim—and reclaim—its own. Contrast such systems with the hundreds of square kilometres of wasteland, usually covered with *alang-alang* grass with which nothing else will grow, which have been created by man in the name of permanent agriculture. Are we so convinced that such agriculture is always the best land use?

'Yes', you will say, that is all very well. But how can we translate the answers small groups of 'primitive' people have found into some that we, the big nations, can use? And even if we could, can we thus hold back the clock, back-pedal on progress? No, there is too much momentum for that, and too many vested interests, which include the family who wants cheap light at night just as much as a government faced with trying to balance better health, education and living standards against its finite resources of oil and timber. But we can at this point in our expansionist type of development pause to think about, to take a long hard look at, what the future will be like if we continue to expand, use up and consume not only our natural resources but our human ones as well.

What will happen to the forest dwellers when there is no more forest? The rain forests of Africa and South America provide, like those of Malesia, a way of life for many of the remaining groups of people whose cultures are different, each one unique, each one with its particular insights into problems often surprisingly similar to our own. Yet hundreds of such tribes or primitive societies have already disappeared from the world. Because their culture depended on oral rather than written traditions, it is gone forever once its practitioners have died out or been absorbed into the fabric of another society. Like any extinction, this represents a loss, in this case to human society as a whole. In social structures, as in natural ones, variety and flexibility mean strength. Yet when, as the anthropologist Margaret Mead has said, we are presented with the results of . . . 'a thousand years' experiment on human beings, we permit the record to be extinguished without protest'.[1] Or, worse still, we destroy it deliberately. In the past, as Mead says, the weapons have been firearms, alcohol and evangelism. Societies have also succumbed to diseases to which they had no immunity, like tuberculosis, venereal disease and smallpox. This is all a matter of historical record, common knowledge, and the aggressive, adventurous Europeans were generally to blame.

But nowadays we are all more enlightened and our social consciences supposedly more awakened. Now we all consider it our duty to spread 'benefits', 'medicine', and 'literacy'. We spread 'living standards', the use of manufactured products with which to achieve them and the concept of labour for money with which to buy them. We take a look at simple people living simply in the jungle and we rush to improve their lot with weapons just as destructive of their way of life as the old ones were, but more subtle and insidious. It is still cultural genocide, and it takes place every day, everywhere a primitive society comes into contact with the irresistible force of a stronger, bigger and more organized state.

In the history of mankind, this process has to be regarded as inevitable: it has always happened, and it will go on happening. No culture can be encapsulated, no living thing can remain alive in a museum. Just as zoos can never substitute for the survival of a healthy population of animals in the wild, or botanic gardens maintain the genetic variety which exists in a rain forest, so we

[1]M. Mead, *Growing Up in New Guinea*, Penguin, 1942.

cannot hope to preserve a dying culture artificially, in a reserve or a park. But we can slow down the pace of the change. With the speed of modern communications and the prevalence of communication devices, to say nothing of our 'hurry' for more raw materials, there seems to be no time anymore for the process of natural change, less painful and dislocating because it is less abrupt. When cultures can evolve from within in a natural response to external change, the process is infinitely less destructive and much that is of value or beautiful in them is retained. This applies to mental attitudes as well as to skills, crafts, art, herbal knowledge, dances, rituals or supernatural powers. One of South-East Asia's most attractive and vital characteristics is the diversity of its people, and the way so many of them have retained, in the face of modernization, or westernization if you like, their own cultural values.

Let us not be too quick to assume that a small group of people with a 'poor' culture has less value than, say, the Minangkabau of Sumatra or the central Javanese. Indonesians will readily think of the Badui people of West Java and of the Batti of eastern Ceram and their reputed powers, or of the Mentawai islanders and the extraordinary sense of balance they have achieved with their rain forest environment. If some restraint is exercised by the modern agents of change, the administrators, the teachers and the loggers, there will still be enough room for such people in the Malesian rain forests—as well as for the plants.

We will probably find that the forest people, who are following their traditional ways of life which use yet conserve the forest, are doing more for civilization than any civilized people are doing for them: man cannot recreate a rain forest once it has been destroyed.

Glossary

Angiosperm : A flowering plant whose seed is enclosed in a fruit.

Aril : Pulpy or thin, partial or complete, covering of a seed. In nutmegs, the mace = the aril. The eye-catching colour and, sometimes, odour of the arils of fruits when ripe are retained by trees dependent on animals for seed dispersal, as is common in rain forest.

Biomass : The mass, by weight or volume, of the living organisms in any one area. A word used for all living matter resident in or rooted in an area. The biomass of the rain forest is greatest at low altitudes, between sea-level and 500 m.

Buttress : A dramatic characteristic of the rain forest, most common in lowland forest, where the tallest trees are also found. Lower stems are enlarged by powerful-looking and often huge modifications of the surface root system. Buttresses can be compared to the stays of a mast, but there is no real evidence that they substitute for a tap root, in the sense of being mutually exclusive. Variations in height, thickness, form and mode of attachment to the tree are distinctive, and help to identify the tree as much as its leaf shape, bark type, sap colour, etc.

Cauliflory : From Latin 'caulis', for 'stalk'. A common characteristic of lowland rain forest is the habit of flowering and fruiting on thickened, old wood (the trunk or thicker branches) rather than at the ends of twigs; or of fruiting at ground level, even below ground. The suggested explanation is that the fruit is thus placed within easy reach of the larger animals who will eat it and disperse the seed.

Flowering and fruiting on the larger limbs or branches is properly called 'ramiflory'.

Climber :   Often called 'liana', climbers have their roots in the ground but, using another plant, usually a tree, for support, succeed in reaching a higher, sunnier part of the forest in order to flower and fruit. They are equipped, depending on the species, with a variety of climbing organs such as hooks or tendrils; their stems are primarily a means of conveying water from root to leaf and flower. Most families of tropical plants contain some climbing species; many have medicinal uses, especially in families in which climbers predominate.

Community :   Any collection or mixture of plants which makes up a distinct type of vegetation. Three factors influence it: climate, soil and other living organisms, which include the influence of plants on each other, and that of insects or animals on the plants. The climatic factors necessary for the existence of the rain forest community only occur in the tropics.

Dicotyledon :   Or 'dicot'; one of the two great divisions of the flowering plants (the other is monocotyledon). It means that the embryo seedling contains two leaves (rather than one). The leaves of dicot plants are nearly always net-veined, i.e., veined both vertically and horizontally in a net-like pattern.

Dioecious :   Having male or female flowers (or cones) on separate plants; this condition ensures outcrossing, i.e., an exchange of genetic material. In the rain forest, where there are many species per hectare and consequently very few individuals of any one species, exchanging genetic material (rather than self-fertilization) is the only way a species can retain its potential to adapt to changing or adverse conditions.

Dispersal :   The transportation or scattering of seeds in the rain forests is rarely by wind, sometimes by gravity and most often by animals. Since fertilization is also usually dependent on animals or insects, rain forest species are far more dependent on both than most temperate climate plants; fortunately there are also many more—and more varieties of— animals and insects in tropical forests than in temperate ones.

Drip-tip : Downward-drooping and sometimes quite elongated tips of the leaves of many rain forest plants enables the leaves to shed surplus water quickly; some leaves are also joined at the base so that they can hang nearly vertically, which has the same effect. In areas of very high rainfall and over 90 per cent humidity, water-shedding is an important ability. However, drip-tips add to the overall similarity of one rain forest leaf to another and have to be more or less ignored when trying to identify a plant.

Epiphyte : A plant which grows on another plant and is not rooted in the soil, but which is not a parasite or climber, its roots being grasping mechanisms rather than nourishment-providers. Epiphytes, e.g., orchids and ferns, occupy a niche in the forest not used by other plants; they add much to the beauty and contribute generally to the appearance of profuse luxuriance of an undisturbed rain forest (see also Myrmecophytes).

Flush and wilt : The young leaves of many rain forest plants are reddish or purple, rather than the usual bright green; they also seem to droop, and sometimes the twig to which they are attached droops as well. Growth modes differ between tree species, and those which display both 'flush' and 'wilt' grow a whole shoot at a time instead of developing their leaves one by one. For lack of chlorophyll, they are pale or reddish the first week, and hang down limp; red colouring in plants is especially common in the mountains where the sunlight is strong, and the pigment may protect the tender leaves. It is also thought that growing a mass of shoots all at once serves to protect at least the majority of tender leaves from insect attack: there are simply too many of them.

Fungus : Fungi (pl.) are a large group of plants which lack chlorophyll and therefore cannot manufacture their own food; instead, they feed on organic matter which they absorb from the material on which they grow via their hyphae. Many fungi are parasites. Fungi play an important role in the rain forest (see Saprophytes and Mycorrhiza) by helping to break down organic material and thus make its food available to other plants.

Gymnosperm : A seed-bearing plant which does not flower. Few families remain of this once large group of plants; they have been superseded, in an evolutionary sense, by the angiosperms. But some of the conifers, for example, are still very successful and are found throughout the world.

Habit : The characteristic way in which a tree grows, i.e., drooping, upright, prostrate, etc., or its form of branching (see Sympodial, Monopodial, and Pagoda).

Habitat : The environment of a plant, its natural living place, to which it is adapted and which supplies the conditions it requires for growth, such as altitude, climate, soil, biological associations (either plant or animal), etc.

Herb : Any non-woody plant whose above-ground parts die off periodically. Most tropical plant families are made up primarily of woody plants—unlike temperate woody plants, which usually have many relatives among the grasses and other soft-stemmed annuals.

Hypha : Filament in the mycelium (vegetative part) of a fungus.

Latex : A viscous white or yellow liquid which oozes from the freshly cut surfaces of many plants. Rubber latex is the best known, but all figs, many climbers, and all but one of the *Artocarpus* genus have this property. Its presence or absence, consistency and colour are good aids in identifying a tree in the forest (see Sap).

Leptocaul : Slender-trunked (of trees), with many branches and slender twigs, as opposed to 'pachycaul'.

Monocotyledon : Or 'monocot'; plants whose embryo seeds contain only a single seed leaf; the other big division of the angiosperms (cf. 'dicotyledon'). Typical 'mono-cot', as opposed to 'dicot', families of plants include grasses, lilies and orchids, all of whose mature leaves have parallel veining rather than net-type veining. Their leaves also form a distinctive sheath around the stem, which, unlike the stem of a 'dicot', attains its final diameter before it lengthens. The basic number of petals and other parts of the flower is often three.

Monoecious : Having unisexual flowers, either male or female on the same plant, e.g., pumpkin vines and *Artocarpus* trees (see Dioecious).

Monopodial : The typical growth form of conifers, for instance, where the main stem (trunk) of the plant goes on growing while producing flowering side branches (see Sympodial).

Mycorrhiza : The symbiotic association between a fungus and the roots of a 'higher', i.e., more complex and evolved plant. Mycorrhizae are either endotrophic, where the fungal hyphae actually penetrate the root cells of the plant, as with most orchids, or ectotrophic, where the fungus covers the finer roots and replaces the usual functions of the root hairs as water- and nourishment-absorbers. Pines, beeches and dipterocarps have ectotrophic mycorrhizae. In general, plants with this sort of assistance are able to make good use of poor soils.

Myrmecophytes : Epiphytes which accommodate ants in their big tubers. Myrmecophytes and other plants inhabited by ants are often found in places where soils are poor, and it is thought that the plant may derive some nourishment from the litter and discarded food brought in by the ants.

Pachycaul : Thick-trunked (of a tree or a palm), with few branches and thick twigs, in the case of a tree. This is the more primitive form of a tree, as postulated by Corner (see Leptocaul).

Pagoda : Special form of either sympodial or monopodial branching, in which the leading shoot grows intermittently, so that apparent layers of leafy side branches are separated from each other by lengths of bare trunk. These trees are always flat-crowned. Also called 'Terminalia branching', named after the *ketapang* or *Terminalia catappa* (the country almond tree) so common on Old World tropical shore lines. As a result of 'Terminalia branching', all the leaves grow in tufts on the upper sides of the branches. Difficult to describe, but very distinctive once seen.

Parasite : A plant that lives on another, taking food from it and giving nothing in return, e.g., many fungi. Plants can also be partially parasitic, in that they may take water (unlike epiphytes which have to store their own) but do manufacture some or most of their own food.

Phenology : The study of the regularly recurring phenomena of plants, such as time of flowering, its frequency, the shedding of leaves, etc.

Pistil : Female organ of a flower, consisting of style, stigma and ovary.

Plank root : A certain kind of buttress root which is very thin or plank-like in vertical section.

Pollination : The transfer of pollen from one flower or plant to another. Self-pollination, in the case of bisexual flowers, can also occur but is usually prevented by different maturation rates of the male and female organs of such flowers. In the rain forest, bats, insects and birds are the main pollinators, rather than the wind which is the usual agent in temperate forests. Bat-pollinated flowers characteristically open in the late afternoon, are white in colour and smell of stale milk. Bird-pollinated flowers tend to be brightly coloured, open in the daytime and have little scent. Flowers pollinated by invertebrates have various shapes, according to whether the pollinator is a wasp, butterfly, bee or beetle. Most are fragrant—in the case of beetle-favoured species, very fragrant; these last also have to be quite 'open' in shape and firm in texture, since beetles are generally large and clumsy.

Ramiflory : Flowering and fruiting on the thicker branches (in

Latin, 'ramus' = branch) instead of on the trunk (see Cauliflory).

Saprophyte : A plant (or other organism) that obtains its food from dead organic matter. Most fungi are saprophytes; so is the long-stemmed creeping orchid, *Galeola*, found on decaying tree trunks. Like all saprophytes, it lacks the green matter, chlorophyll, which would enable it to synthesize its own food from the sun.

Saps, resins and gums : Juice, viscous liquid, resinous liquid or latex found in the outer bark, twigs, leaves, or other parts of plants such as unripe fruits. Resin canals occur throughout the wood of most dipterocarps. 'By their sap ye shall know them'; well, almost. 'Slash', or the botanists' knife test helps to identify at least the family. The wound oozes a red blood-like sap in the nutmegs; white and sticky in most *Artocarpus* plants and figs; transparent or milky and glassy hard after exposure to air for a while in the *Agathis*; from pines, the resin is made into turpentine. Various dipterocarps can be tapped for *minyak keruing* (*keruing* oil) and *damar*; clear *damar* from the dipterocarp, *Shorea javanica*, the *damar mata kucing*, 'cats' eyes', is said to be the most valuable and the same term is used for clear *Agathis dammara*.

Shifting cultivation : A form of agriculture in which fields are rotated instead of crops, often practised by so-called primitive peoples; its success presupposes low population density.

Stamen : Male part of a flower which carries the pollen.

Stenophyllism : The tendency towards a markedly narrow leaf shape, often displayed by rain forest trees growing near rivers.

Stilt root : Stilt roots grow out from the lower part of the stem and arch outwards and downwards like thin aerial buttresses common in swamp forests, on mangrove plants and pandanus (see Buttress).

Succession : Plant re-colonization of open ground in a certain order—those which can stand the most heat and light first. The order of succession of rain forest genera from first 'weed' through *'belukar'* (secondary forest) to true 'climax' or 'primary' rain forest vegetation is little known. Richards (see Bibliography) calls this 'the most serious gap in our present knowledge of the rain forest' in these days of widespread and rapid deforestation.

Symbiosis : An intimate association between two organisms of different species to their mutual—but not always equal—benefit.

Sympodial : A form of branching in which the main stem of the plant dies away periodically or ends in a flower, while the growth is continued by side shoots, e.g., *ketapang*, or country almond, *Terminalia catappa* (see Monopodial).

Bibliography

Ashton, P. S. (1976), 'Ecology and the Durian Theory', *Gardens Bulletin*, XXIX: 19–23, Singapore.

Beekman, E. M. (ed. and trans.) (1981), *The Poison Tree: Selected Writings of Rumphius on the Natural History of the Indies*, Amhurst: University of Massachussetts Press.

Brown, R. (1821), 'Account of a New Genus of Plants named Rafflesia', paper read to the Linnean Society, 30 June 1820, *Transactions of the Linnean Society*, XIII, 201–34, London.

Burkill, I. H. (1935), *Dictionary of the Economic Products of the Malay Peninsula*, London: Crown Agents; reprinted Singapore: Governments of Malaysia and Singapore, 1966.

Burley, J. and Styles, B. T. (eds.) (1976), *Tropical Trees: Variation, Breeding and Conservation*, London: Academic Press.

Colchester, M. (1989), *Pirates, Poachers and Squatters: The Dispossession of the Native Peoples of Sarawak*, Kuala Lumpur: Survival International/INSAN.

Collins, M. (ed.) (1990), *The Last Rain Forests*, London: Mitchell Beazley.

Cooling, E. N. G. (1968), *Pinus merkusii*, Commonwealth Forestry Papers No. 4, Oxford.

Corner, E. J. H. (1940), *Wayside Trees of Malaya*, 2 vols., Singapore: Government Printing Office; reprinted Kuala Lumpur: Malayan Nature Society, 1988.

_____ (1949), 'The Durian Theory and the Origin of the Modern Tree', *Annals of Botany*, NS, XVII, No. 52 (October).

_____ (1964), *The Life of Plants*, London: Weidenfeld & Nicholson.

_____ (1966), *The Natural History of Palms*, London: Weidenfeld & Nicholson.

Cranbrook, Earl of (1988), *Key Environments: Malaysia*, Oxford: Pergamon Press, and Gland, Switzerland: IUCN.

Cribb, R. (1988), *The Politics of Environmental Protection in Indonesia*, Centre of South East Asian Studies Working Paper 48, Monash University.

Denslow, J. S. and Padoch, C. (eds.) (1988), *People of the Tropical Rain Forests*, Los Angeles: UCLA Press/Smithsonian.

Dransfield, J. (1974), *A Short Guide to Rattans*, Biotrop Monograph No. 7, Bogor.

Heyne, K. (1950), *De Nuttige Planten van Indonesië*, 's-Gravenhage: van Hoeve; first published 1927.

Holm-Nielsen, L. B., Nielsen, I. C., and Balsev, H. (eds.) (1988), *Tropical Forests: Botanical Dynamics, Speciation and Diversity*, London: Academic Press/New York: Harcourt Brace Jovanovich.

Jacobs, M. (1976), 'The Study of Lianas', *Flora Malesiana Bulletin*, 29: 2610–17, Rijksherbarium, Leiden.

———— (1987), *The Tropical Rain Forest: A First Encounter*, English edition of his 1981 *Het Tropische Regenwoud*, Remke Kruk et al. (eds.), published posthumously by Springer-Verlag, Berlin.

Janzen, D. H. (1974), 'Tropical Blackwater Rivers, Animals and Mast Fruiting by the Dipterocarpaceae', *Biotropica*, 6: 69–103.

———— (1986), 'The Future of Tropical Ecology', *Annual Review of Ecology and Systematics*, 17: 305–24.

Leighton, M. and Wirawan, N. (1986), 'Catastrophic Drought and Fire in Borneo. Tropical Rain Forest in Association with the 1982–83 El Niño Southern Oscillation Event,' in G. T. Prance (ed.), *Tropical Rain Forests and the World Atmosphere*, A.A.A.S. Selected Symposium 101, Boulder, Colorado: Westview Press.

Leith, H. and Werger, M. J. A. (eds.) (1989), '14A Tropical Rain Forest Ecosystems, Structure and Function'; '14B, Tropical Rain Forest Ecosystems, Biological and Ecological Studies', in *Ecosystems of the World*, Amsterdam: Elsevier Scientific Publishing Co.

Mackie, C. (1984), 'Lessons Behind East Kalimantan's Forest Fires', *Borneo Research Bulletin*, 16: 36–74.

Marshall, A. G. (1983), 'Bats, Flowers and Fruits: Evolutionary Relationships in the Old World, *Biological Journal of the Linnean Society*, XX: 155–85.

Myers, N. (1979), *The Sinking Ark*, Oxford: Pergamon Press.

———— (1984), *The Primary Source: Tropical Forests and Our Future*, New York and London: W. W. Norton & Co.

Prescott-Allen, R. and Prescott-Allen, C. (1982), *What's Wildlife Worth?* London: 11ED/US: WWF.

———— (1983), *Genes from the Wild*, London: Earthscan.

Quisumbing, E. (1978), *Medicinal Plants of the Philippines*, reprint of 1951 edition, Quezon City.

Piper, J. M. (1989), *Fruits of South-East Asia*, Singapore: Oxford University Press.

———— (1992), *Bamboo and Rattan*, Singapore: Oxford University Press.

Repetto, R. and Gillis, M. (1988), *Public Policies and the Misuse of Forest Resources*, Cambridge: Cambridge University Press.

Richards, P. W. (1952), *The Tropical Rain Forest*, Cambridge: Cambridge University Press.

Rubeli, K. (1986), *Tropical Rain Forest in South East Asia: A Pictorial Journey*, Kuala Lumpur: Tropical Press Sdn. Bhd.

Silcock, Lisa (ed.) (1989), *The Rain Forests: A Celebration*, London: Living Earth Foundation, Barrie & Jenkins.

Spencer, J. E. (1977), *Shifting Cultivation in Southeastern Asia*, Berkeley: University of California Press.

Whitmore, T. C. (1984), *Tropical Rain Forests of the Far East*, 2nd edn., Oxford: Clarendon Press.

———— (1990), *An Introduction to Tropical Rain Forests*, Oxford: Oxford University Press.

Woods, Paul (1989), 'Effects of Logging, Drought and Fire on Structure and Composition of Tropical Forests in Sabah, Malaysia', *Biotropica*, 21/4 (December): 290–302.

Journals/Newsletters, with particular reference to South-East Asia

Flora Malesiana Bulletin, J. F. Veldkamp (ed.), Rijksherbarium, Leiden. News, reviews, articles. Conservation update in each issue, by H. P. Nooteboom.

Malayan Naturalist, published by the Malayan Nature Society, Kuala Lumpur.

Newsletter of the Association of South East Asian Studies in the UK. First of series, spring 1987, Centre for South-East Asian Studies, University of Hull.

Oryx Journal of the Flora and Fauna Preservation Society, UK, Blackwell Scientific Publications, Oxford.

Royal Society's South East Asian Rain Forest Research Programme Newsletter. 1st issued 1985. A. G. Marshall (ed.), Department of Zoology, Aberdeen University, Scotland.

Wallaceana, official newsletter of the working group of tropical ecology. (INTECOL) Phang Siew-Moi (ed.), Institute for Advanced Studies, University of Malaya, Kuala Lumpur.

Index